THE OILMEN
THE NORTH SEA TIGERS

Plates 1 and 115.

First and last pages. The oilmen – some of the crew members of CNR's Murchison platform in the North Sea, West of Shetland. The cameraman, Sean Newman, has been a production technician on the platform for ten years. Some of these men have now retired. Sean had seen an exhibition on the oil industry in Aberdeen's Maritime Museum and was struck by the lack of pictures of people. So he offered the musuem his remarkable series of portraits.*(Photographs: Sean Newman. Montage: Greybardesign)*

BILL MACKIE

THE OILMEN
THE NORTH SEA TIGERS

Birlinn

First published in 2004 by
Birlinn Limited
West Newington House
10 Newington Road
Edinburgh
EH9 1QS

www.birlinn.co.uk

Copyright © Bill Mackie, 2004

The right of Bill Mackie to be identified as the
author of this work has been asserted by him
in accordance with the Copyright, Designs
and Patents Act 1988

All rights reserved. No part of this publication
may be reproduced, stored, or transmitted in any
form, or by any means, electronic, mechanical or
photocopying, recording or otherwise, without
the express written permission of the publisher

ISBN 1 84158 302 2

British Library Cataloguing-in-Publication Data
A catalogue record for this book is available
from the British Library

Layout and page make up: Mark Blackadder

Printed and bound by GraphyCems, Spain

Contents

	ACKNOWLEDGEMENTS	7
	PICTURE CREDITS	9
	PREFACE	11
1.	The New Oil Age	13
2.	The Pioneers	25
3.	Life Offshore	46
4.	The Good, the Bad – and the 'Coonasses'	70
5.	A Breed Apart	83
6.	The Supporting Cast	103
7.	First Oil Flows	132
8.	Lessons from the Past	162
9.	Who Listens to the Piper?	181
10.	Human Interest or Business Interests?	195
11.	Not Required Back	210
12.	Women and Home	223
13.	Camaraderie and the Crack	237
	EPILOGUE	252
	APPENDICES	259
	GLOSSARY	261

To my dear wife, Dorothy.

Acknowledgements

This book began germinating late one summer afternoon forty years ago, when a young newspaper features writer boarded a small frail boat which had sailed from the Gulf of Mexico to a mooring at Regent Quay, Aberdeen. The tall story I was told by a lanky laconic Texan geologist about what was reputed to lie beneath the grey waters beyond the harbour bar stimulated a lifelong obsession with the development of what was to become the fabulous North Sea oil and gas industry. Over the years it would be true to say, therefore, that many hundreds of people have contributed to my knowledge and awareness of the world offshore and thus to the writing of this book. My first attempt at analysing what they told me resulted in a doctoral thesis, *The impact of North Sea oil on the North-east of Scotland, 1969-2000* (The University of Aberdeen, 2001). The research and some of the interviews provided the basis for elements of *The Oilmen* and I must thank again all those who assisted me in completing that enterprise. It is unfair to single out people this time, but I am particularly grateful to Ernie Wight, who unearthed some marvellous photographs and introduced me to the archives of the unsurpassed company newspaper, *BP Oil Producer*; Keith Webster, formerly of Conoco, for finding me superlative American interviewees; as always, my oil industry 'guru', Dr Colin Webster; among the pioneers, Joe Dobbs and his irrepressible mate, Swede Lingard, Bob Balls, the redoubtable Kevin Topham, Brian Porter, Ralph Stokes, Michel Euillet, Graeme Paterson, Mike Waller, Caroline and Steve Russell-Pryce, Dick Winchester and Captain John Carter; from Piper Alpha, the late Bob Ballantyne, Mike Jennings and Ann Gillanders; from the oil companies, Ruth Mitchell of Chevron Texaco, Jack Page, Caroline Kinghorn of Schlumberger and Richard Grant of BP; John Wils; Jake Molloy and Dave and Lorna Robertson of the OILC and Ronnie McDonald and my good friend, the late George Watt. But my thanks go to all of the representatives of generations of oil workers – men and women – who freely gave of their time to relive their experiences. One and all they are extraordinary folk who, across the tumultous final decades of the twentieth century, against all the odds and too often at great human cost, helped to build a new world out where there was nothing. Theirs' has been a tale well worth the telling. They have my unbounded admiration.

Bill Mackie
Aberdeen
September 2004

Picture Credits

The main credit for the collation of most of the images must go to the tireless picture researcher, Christine Maclean. The brilliant montages and graphics are the work of my good friends at Greybardesign. I would like to thank the following for giving me permission to use their copyright photographs: Jim Fitzpatrick who has an amazing comprehensive photographic history of the North Sea industry; Kate Sutherland; Allan Wright; Sean Newman; Kenny Thompson; and a too numerous to name band of oilmen who allowed me to plunder their personal archives. I am also grateful to the following for the use of their archival material: Bond Helicopters, Bristow Helicopters, CHC Scotia, CNR International, North Star, Sea Lion Shipping, Subsea 7, BP, Shell and Conoco Phillips; *Grimsby Evening Telegraph* and *Nottingham Evening Post*.

Some people find oil, some don't

JEAN PAUL GETTY

We were given the centre cut of the watermelon

JACK MARSHALL,
AMERICAN OILMAN

North Sea oil is God's last chance for the British

SIR ANDREW GILCHRIST,
CHAIRMAN HIGHLANDS AND ISLANDS
DEVELOPMENT BOARD

Preface

North Sea oil and gas is the great industrial triumph of the twentieth century and the largest, most successful, single enterprise in the modern economic history of Scotland.

Massive in concept, in employment and in the generation of vast wealth, the harvesting of hydrocarbons from some of the most unforgiving waters on Earth has been a triumph of pure human endeavour to equal any achievements in space.

The fiscal revenues saved a debt-ridden United Kingdom. Vulnerable Scottish communities have been energised and enriched. New opportunities and new companies have emerged and from the golden stream of profits, global giants of commerce have grown.

Out of nothing, out of the empty spaces of offshore Britain and in a brief span of time, thousands of pioneers have created a new world. Bold, brave, imaginative and innovative – they crossed an unknown frontier. Driven by tough, demanding masters they risked their lives and too often, too many died.

This is a tribute to the offshore work force of the North Sea. These are some of their stories.

The New Oil Age

Early October 1970, on the drill floor of the semi-submersible *Sea Quest*, located on block 21/10–1 in the Central North Sea, 130 miles off Aberdeen, on the east coast of Scotland.

'"Mick the Wad" Waddington, he was the driller. I was the derrickman and the rest of the crew the same as always,' recalls eighteen-year-old farm boy, 'Swede' Lingard. 'We kept drilling and the oil must've been seeping into the mud 'cos when it got to the surface it was coming out on the shakers. I kept going up to Mick the Wad and saying, "I reckon the bluidy rotary table seal's gone." I thought it was lubricating oil coming out of the table. We got into it and we knew we had a drilling break, but Mick, he said, "No, there's just different sand here and we may have to go down another couple of hundred feet deep." We had Corelab doing the samples and we had a BP geologist looking at them. Mick said we'd better stop. That ol' geologist from Corelab, he said, "Bluidy 'ell, we've got some sand with some shale in it." Oh, right, right, right, I thought.' Thirty-four years on, Swede has never forgotten that moment when the crew of the *Sea Quest* became aware that something significant had happened.

'What a day that was,' said the company's senior geologist, Laurie Horobin, who examined the first and only core extracted from a few feet above the reservoir on block 21/10–1. 'I distinctly remember the core – and there was only one, a column of almost pure white sand. Any trace of oil appeared to have been bleached out of it but there was just a whiff off it. Mind you, that didn't always mean anything.' *Sea Quest* had been hunting in the empty acreages of the North Sea in a cluster of blocks awarded to BP in the second licensing round in 1965 and the company were not optimistic. According to regional geologist, Peter Walmsley, in the anthology *Tales from Early UK Oil Exploration 1960–1979*,* only some promising results from earlier geophysical mapping up to 1970 'had persuaded the general

Plate 2.
British Petroleum's drilling rig, *Sea Quest*, queen of the North Sea oil and gas discoveries. The SEDCO semi-submersible launched the multi-billion UK industry by finding the fabulous Forties Field. (BP plc)

---

* Moreton, R., ed., *Tales from Early UK Oil Exploration 1960–1979*, Petroleum Exploration Society of Great Britain 30th Anniversary Book (London, 1995).

Plate 3.
Operating the tongs on the drill floor on *Sea Quest*, somewhere in the North Sea. Roughnecks Little Joe Dobbs, from Grimsby, Gordon Lawson, from Aberdeen and Gordon Miller from Sheffield.
(BP plc)

management of a rather impoverished BP to drill without delay and not to farm the blocks out as they had been contemplating'. In fact there was pressure internally to sell their acreage, with the likely beneficiaries the rival Shell oil company and their American partners, Esso, who were drilling fruitlessly further north. And the clock was ticking towards the five-year time limit imposed under the licensing concession. 'Commitments made at this time,' said Walmsley, 'were very much an act of faith or a shot in the dark.' Six months previously, BP chairman Sir Eric Drake had made the now notorious pronouncement, 'There won't be a major field in the Northern Sea.' Veteran international oil consultant and broker Alex Barnard is adamant this was not just a typical oilman's commercial subterfuge. 'He said he was only following advice given by his senior geologists and he had regretted it ever since.'

Security on *Sea Quest* had been obsessive and paramount from the beginning. 'All drilling and geological messages were scrambled or transmitted in code,' said Walmsley, 'except one.' This was a phone call to his home at about 4 a.m. 'I will always remember the guarded words, "Peter, it looks good." "That's fine," I said, "thanks for calling," and contentedly went back to sleep.' While the well was being tested, a senior BP manager flew out to the rig and imposed a total security blackout, erecting temporary

Plate 4.
A fortune in a bottle – a sample of the first crude oil from the field that was to be named Forties.
(BP plc)

walls round the drill floor. Swede Lingard was still indignant at the recollection. 'He weren't popular, that feller. He wouldn't let us off because he didn't want anyone talking about it. I think it was two or three days – and we'd only got seven days off. He weren't popular at the best of times. He could pick a fight in an empty room, but that wasn't on.'

But the operational blackout was crucial, as Laurie Horobin recalls. 'Everything was under wraps for quite a long time. I don't think anybody was sworn to secrecy, but there were quite a lot of restraints on people — and I think the phones were cut off for 24 hours. Mind you, it was pretty obvious very quickly that something was happening. You daren't do a thing because not only would it affect the stock markets, but there were other concessions up for grabs adjacent to Forties. When that oil come out – it was very green and light and it just smelled like and looked like creosote. We flow tested it. BP were trying to keep it quiet, but with all the trawlers coming and going into town a lot of rumours were flying around Aberdeen.'

It was not until 1971, when the values of the field had been proved and negotiations were underway for the land the pipeline was to be built along, that BP announced the discovery to the world. Television journalist Ted Brocklebank has a clear memory of that moment. With other media representatives, he had been primed to meet a helicopter arriving at Dyce Airport. 'A

THE OILMEN

man in a hard hat and a tartan shirt came down the steps of the helicopter and he held up what another reporter said looked like a salad cream bottle with flat Guinness in it. The man said, "Gentlemen, this, is North Sea oil." We didn't really appreciate what it all meant until a couple of days later we were taken out to *Sea Quest*. Then we were told the facts and it began to sink in.'

The Buchan farmer, Maitland Mackie, who was the convener of Aberdeenshire County Council, first heard the news at a meeting in London. He could still remember the occasion twenty-five years later. 'When the BP fella told us the size of the field there was just a stunned silence that lasted for ages. And then he said – this was McLeod Matthews, the PR man – that they reckoned it would cost £370 million to develop it.' Even the canny businessman, with connections to Texas through his American wife, was taken aback. And he was one of the few people in the North-east of Scotland who quickly recognised the potential for his native area in these secretive expeditions into the North Sea.

Swede Lingard was more matter-of-fact. 'The crew, we all took samples of the stuff. Mine's in a horseradish bottle somewheres.' That bottle of oil, the first from the 1.8 billion barrels of strategically invaluable reserves in what was to be BP's fabled Forties Field, is symbolic of epic human achievement, the signal for the birth of the North Sea oil industry. To envisage the sheer scale of it is hard even today in this highly technological age when the daily darg of the industry to keep the oil – and the revenues – flowing has become a commonplace. It took the worst disaster in the history of the oil industry, the destruction of the huge Piper Alpha production platform in July 1988 and the deaths of 167 men, to finally force into perspective universal recognition of the achievements of the oil and gas work force and of the latent power of the elements that they had harnessed, but not tamed.

The thirty-five-year history of North Sea oil is crowned with more superlatives than that of any other modern industry. It was a logistical exercise as massive and complex as the wartime Normandy landings: calculating workers in their thousands, barrels of oil in millions, capital expenditure in thousands of millions; building dozens of iron and concrete structures higher than St Paul's Cathedral and planting them hundreds of miles out in the treacherous North Sea; and laying massive lines of pipes to carry the oil and gas across the seabed and thence through Scotland. To call it pioneering frontier work is not hyperbole. The oilmen were butting impatiently at the leading edge of marine engineering and 'inner space' was as difficult and dangerous to conquer as outer space. Coincidentally, man landed on the moon in the same year the first oil was discovered.

In truth, the modern hunt for marine hydrocarbons and the exploitation of the complex, rich basins of oil and gas off the shores of the United Kingdom had many beginnings. For 150 million years, reservoirs of hydrocarbons cooked from billions of fragments of organic matter by the intense heat and pressure had accumulated deep in hidden sedimentary basins and crevices below the seabed of the North Sea. Onshore, their existence had been known down the ages as oil seepages and put to a variety of uses; oil-soaked peat in Lancashire made effective firelighters, while coal miners lubricated their cartwheels from the substances oozing out of rocks in the pits. Offshore, awareness of the presence of fossil fuel was, however, of a more modern provenance. In 1854, the geologist Hugh Miller of Cromarty, wrote of the Moray Firth, 'Every heavier storm from the sea, tells of its [oil's] existence by tossing ashore fragments of dark bituminous shale. The shale is so largely charged with inflammable matter as to burn with a strong flame as if steeped in tar of oil.' The same sedimentary rock became the centre of what is generally accepted as the birth of the modern commercial oil industry when the chemist James Young perfected refining oil from shale at a plant near Bathgate in West Lothian in 1850. By the end of the century, the industry he founded, which had spread to Fife, was producing 2 million tons of oil a year. In comparison, this was less than the North Sea oil industry could produce in a week. The shale business became uneconomic and petered out in the early 1960s.

At sea, Scots fishermen, instinctively in harmony with, but ever wary of, the moods of their treacherous catching grounds, knew all about marine oil, according to Bill Adams, former North-east officer of the Scottish Council for Development and Industry. He grew up in the little fishing port of Stonehaven where a trawl company manager once told him, 'Fishermen have always known where the seepages of oil were by the gas bubbles on the surface of the sea. They regularly brought up oil tar balls in their nets.' Myles Bowen, a Shell geologist, acknowledged later when surveys failed to turn up evidence of hydrocarbons, 'If anyone had bothered to consult the North Sea fishermen they might have discovered otherwise.'

The long, laborious route to the North Sea began in the late nineteenth century when an American colonel, who was looking for a source of salt, struck Spindletop, the world's first oil gusher, on the banks of a creek in Pennsylvania and produced the first commercially viable fossil fuel. A host of sub-surface wells quickly followed in America and across the world as the development of the internal combustion engine in the 1870s provided the catalyst for a huge increase in global oil production. Demand was accelerated

THE OILMEN

by the mechanisation of warfare during World War One. The next incentive for the industry came from the tireless assembly lines of the rapacious new automobiles, forcing the companies to widen their search for new locations. The Americans spread out across their home states and into South America, where a British company, Shell Transport and Trading, which had first struck oil in Borneo, also investigated prospects in Venezuela. Among the other British oil interests which ranged throughout the Far and Middle East was a company assembled by a determined wealthy English speculator called William Knox D'Arcy, who was financed by Burmah Oil, a Scots concern based in Glasgow. After seven difficult years the D'Arcy Exploration Company found a ten-square-mile oilfield in the deserts of Persia. The company then became part of Anglo-Persian Oil while Burmah's stake was bought out by the British Government who were looking for secure supplies for the fleet, on the orders of First Lord of the Admiralty, Winston Churchill.

By the 1930s, promising geological formations on British soil were attracting interest, and in 1939 came the first tentative native foray into the oil business. D'Arcy, now the drilling arm of Anglo-Iranian, brought in the country's first commercial onshore oilfield at Eakring, a small Nottinghamshire village in Sherwood Forest, at the heart of England. It became the unlikely focus of a thriving land-based industry reaching into Lincolnshire and Dorset. Oil and the oil companies by then were dominating the world energy market both in the use of hydrocarbons as the primary power source and through new chemical by-products. Stimulated by a second global conflict, consumption grew at a phenomenal rate – in Western Europe alone, demand increased fifteen-fold between 1949 and 1972. The controlling cartel of five major American and two British oil companies, dubbed 'The Seven Sisters', were tirelessly scouring the globe for new sources.

In Europe, the next leap forward was the discovery in 1959 by Shell and Esso of a vast gas field in the coastal marshes of Holland. The reserves of Groningen, the most voluminous outwith the USSR, were reckoned to be so enormous they could fulfil the gas demands of the UK for a hundred years. They also made the country hugely wealthy and self-sufficient. Intriguingly, the proving of the Slochteren No 1 well seemed to confirm a long-held suspicion that beneath the bed of the North Sea, identical rock formations might be a continuation of the land geological structures already yielding hydrocarbons for the Netherlands' contiguous coastal neighbours, Britain and Germany. New thinking had been emerging since the 1950s about the formation of the rock layers beneath the sea, based on the involvement of plate tectonics in the process known as continental drift. This had caused the deposit of layers of porous rock, known as sedimentary basins in the area,

Plate 5.
Land drilling operations onshore in the UK at the Nottinghamshire oil fields at Eakring. The first oil was struck in 1939 by the D'Arcy Oil Company, later to become BP. *(BP plc)*

which became the North Sea. Certain other geological events combined with pressures on the land mass had caused faulting, resulting in the creation of traps, which could possibly contain oil and gas. This was the theory petroleum geologists were eager to exploit. The barrier, however, was finance. Peter Hinde claimed in the PESGB collection, *Tales from Early UK Oil Exploration 1960–1979*, 'There was little encouragement towards the great expense of developing an oilfield under the sea which cost four times that of a land operation.' There was also the question of the capabilities of the explorers and their equipment. In the same anthology, geologist Leslie Illing maintained the industry was in its infancy. 'Several years were to pass before the necessary marine drilling rigs capable of operating in the decidedly hostile environment of the North Sea were to become available.' Yet the prospect of operating in a politically stable country, which shared a common

THE OILMEN

cultural heritage and language and had a proven history of international alliances and mutual co-operation, was highly seductive to the mainly American oil companies, who were frustrated by the endless power struggles with new, volatile oil nations which were fully aware of their bargaining strengths.

Oilmen had already ventured successfully offshore. First under the Persian Gulf, drilling a primary well as early as 1908, which wasn't developed until the 1930s. The province of fabulously wealthy Arab states that grew from that first discovery was so prolific it still represents 50 per cent of the world's total oil reserves. By then the first marine drilling had begun in shallow coastal waters and in lakes in the Caspian Sea and in Lake Maracaibo, in Venezuela, where it was apparent that oil deposits extended under the water. These first operations in the 1920s had to be carried out by barges and then from platforms fixed to the seabed. Both areas continue to attract intense activity by the international oil industry. Then, in 1947, drilling company Kerr-McGee began the American offshore industry by bringing in the first well 'out of sight of land' in the Gulf of Mexico. So by the time the North Sea was ready for exploitation, marine techniques were well established. Another conclusive imperative for security of supply in Europe was the Suez crisis of 1956 and the rebirth of Arab nationalism, imperilling the West's sources of oil.

The door to the North Sea was finally thrown open in 1964 through two key political developments. Britain ratified the Geneva Convention of 1958, which established national jurisdictions offshore and exposed the Continental shelves of individual countries worldwide to exploration. The turbulent North Sea, replete with hidden natural wealth, was divided by a meridian line into equal territories between the UK and Norway. That year, the British Government fired the starting gun for the most dramatic industrial and commercial enterprise in the island kingdom's history. Nine hundred and sixty blocks or parcels of areas, representing the bulk of the designated UK North Sea sector from latitude 52 to 62 degrees, were offered for exploration and development under licence under the Continental Shelf Act. Twenty-two consortia, representing fifty-one companies, were awarded fifty-three licences for 394 blocks. The southern North Sea was most heavily favoured because of the contiguous coastal geology and the onshore gas finds. The prospect of oil was not then thought likely, and only a few of the bolder gamblers chose sections of the northern area for their geologists and seismologists to survey and prospect. The majority of the licence holders were based in the United States; only 30 per cent were British. Among them was the former Anglo-Iranian Company, which in 1954 had been renamed the British Petroleum

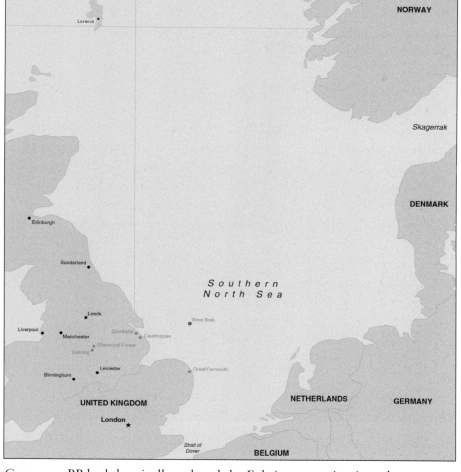

Figure 1.
The southern North Sea sector in the mid-1960s with the first commercial gas find by BP on what was to become the West Sole Field. The oil companies operated out of bases at Cleethorpe and Grimsby.
(Greybardesign)

Company. BP had drastically reduced the Eakring operation in order to divert the thrust of their exploration drive from onshore to offshore.

The global focus was now on the North Sea. Throughout 1964, the first marine drilling rigs to work in Europe were determined to replicate the good fortune of the Dutch and to find gas across an area that eventually widened from Teeside in the north to Norfolk in the south, where onshore activity was to centre mainly on the fishing port of Great Yarmouth. BP also had a base at Cleethorpes in Lincolnshire. The drillers were armed with geological maps, charts and data provided by the seismology teams, the advance scouts of the industry. But the opening sorties were not encouraging. The first exploration well, drilled for Standard Oil of California and Texaco by *Mr Cap*, a semi-submersible, was dry. Success finally came in the following year, when BP's *Sea Gem* jack-up drilling barge struck gas, the first well of the West Sole field. This opened up of what became the Southern Gas Basin, whose flow of gas ultimately made Britain domestically self-sufficient in the

THE OILMEN

Plate 6.
A Santa Fe crewman looks back at his workplace, Platform A on the West Sole Field in 1966. Fog had grounded the normal helicopter air ferry and the crew change had to be made by boat. *(BP plc)*

fuel that supplanted coal in powering homes and industry. But the triumph was later marred by tragedy. On Boxing Day 1965, the *Sea Gem* sank with the loss of thirteen crew members. It was the first major disaster of the North Sea industry and unfortunately not the last (see Chapter 8).

The rig that took over the task of completing the West Sole well was the new BP semi-submersible, *Sea Quest,* which came into service in 1967. The £3.5-million American-designed drilling platform was built at the Harland and Wolf shipyard in Belfast. At 7,500 tons, she was one of the largest offshore rigs ever constructed, dwarfing three slipways in the yard, each of her three legs 35 feet in diameter. A giantess, created by the South Eastern Drilling Company (SEDCO) to withstand the rigours of the North Sea, she was the third in the company's first series, the 135s. Canadian Bob Archison joined SEDCO in 1965, chosen for his experience – as a railroad mechanic. The rationale was that offshore drilling rigs were originally powered by huge locomotive-style diesel engines. Bob explained that the early rigs were given the number 135 to indicate that they could sit on the bottom in 135 feet of water. The company, which eventually became the largest driller in the new province with a peak of fourteen rigs, now belongs to Transocean.

Among the crew supplied by the drilling contractor, Santa Fe, on *Sea Quest* were two young raw recruits, Rob 'Swede' Lingard, fresh from his father's farm and 'Little' Joe Dobbs, who had been a hotel *commis* waiter. Little Joe remembers they had to clear a lot of debris from the well. Swede saw a couple of the *Sea Gem*'s legs sticking out of the water. 'They were actually blowing them up with dynamite while we was working on the well!' The *Sea Quest* confirmed West Sole A before moving on to look at six or seven other locations.

One night they were drilling off Whitby when the unpredictable North Sea struck. The irrepressible Swede provides this frightening recollection. 'I think it was 72 feet this wave – you used to get one every hundred years, they said, and it came crashing through the rig. Because it had three legs it used to roll like a cork and it dipped down at the front. I was on the rig floor and all I could see was water all down the front of the rig. The wave hit the side of the helideck and come over straight through the middle of the rig. The bluidy riser was snapped clean in half just under the rotary table and the whole thing, about 200 feet of it, just went flying and it landed on the sea bed about 40 feet away. That wave bust bluidy winch lines and everything. Harry Miller was the toolpusher. "Harry the Fist" they used to call him – rough as soot he was. He says, "We'll have to get cracking. We're going to have to let ourselves go." We were frightened we were going to break something.

'Harry says the tugs are on the way, out of the Humber. "What we're going to do is cut the remaining cables." Harry said, "I'll cut them," and I said, "Harry, I'll watch you." But I didn't. I said, "I'll cut one side and you cut the other." So we got ourselves wrapped round the handrails with a cutting torch strapped to a broomstick. We was leaning back about 5 or 6 feet and the cables were 4 inches thick. When they went, by hell, didn't they bluidy go. Wire flew everywhere. It went over the top of the winch and took about 20 feet of handrail off, but we were below them out of the way. But by heck – it didn't half make things rattle. The rig took off then, bobbing about on the waves just like a piece of wood. Harry, he said, "We'll have to slow down a bit." We're doing something like 6 or 7 knots 'cause we was out of the water by then. But the tugs were only doing 4, so they couldn't catch us.' As the drilling rig drifted further and further south, the crew began to worry about the proximity of the Dogger Bank.

'Says Harry, "We'll have to put some drill collars over the side on a big crane to slow us down." I said to Harry, "I think it'll pull the crane off its pedestal." Luckily, we talked him out of it. Then we got a bit of a lull in the weather and we turned. The weather came back up again and the tugs by

THE OILMEN

Plate 7.
Restored to working order after a terrifying breakaway across the stormy southern North Sea, *Sea Quest* allows a small flotilla of tugs to take her north across the 56th parallel in 1968 and towards the most famous discovery of all. *(BP plc)*

this time were quite close. Harry said, "We'll have to shoot a line with a Verey pistol on to a tug and fix up a hawser." He went, "You're a shooting man?" "Oh, yes," I said. "Right then," he said, "see if you can land a line on the back of that boat." Well, I shot this bluidy pistol and the rocket went straight through the wheelhouse window. I said, "Is that near enough?" That rocket went round that bluidy bridge for about 5 minutes and the crew were diving out of the way. But we got the line on. Next day the tug captain said, "Next time, tell him to shoot it at the deck, not at the bluidy wheelhouse." It took us about 5 hours but we got tied on. That all happened over about three days. I have had hairier moments but it was all a good laugh at the end of the day.'

There was a sequel. The crew of forty were eventually taken off and *Sea Quest* was towed on to a sandbank in Bridlington Bay. 'Next day a load of welders come out from Grimsby docks and did temporary repairs to the cruciform at the bottom of the legs. They had cracks you could put your penknife in.' The *Sea Quest* then put into Rotterdam for four months for repairs and modifications. The southern sector of the North Sea was not quite finished with the drillers, however. They had to return to the well to abandon it and pick up the blow-out preventer from the seabed. But a trawler got there before them and hit the BOP. 'We spent weeks and weeks with divers picking up the risers. There was a 40-ton drill down there we never did find, it must have sunk in the mud. But we cemented the well and made it safe. Then we drilled a few more wells, six I think it was, and then we more or less moved up to Aberdeen – looking for oil.' That is when the story of oilmen like Swede and Little Joe and all the other pioneers of the most valuable section of the North Sea really began.

The Pioneers [2]

The image of the North Sea as a place of beauty runs contrary to the reality of the sullen icy grey waters that, since the first human settlements, have so defined life for the fringe of communities around the east coast of Scotland. Yet those were the terms in which French diver Michel Euillet chose to describe one winter's day offshore, in 1973. 'I remember we had snow. Now, this is quite beautiful on a rig. You feel as if you are inside one of those glass balls you shake to make a snowstorm. All noise is cut out so you don't even hear the drilling. You have become so used to the noise, but in the snow there is silence. And if it is Christmas and a helicopter is landing, you feel that Father Christmas is going to arrive. Unfortunately it is just a crew change. The romance is all in your head but if you don't have it during a 12-hour shift – you would go mad.'

The northern North Sea oil and gas industry was already four years old when Michel enjoyed that experience. It is unlikely, however, that the oilmen who left the southern gas basin in their drilling rigs early in the winter of 1967 and set a course north over the 56th parallel deep across unknown waters, would have been harbouring such aesthetic notions. From the outset the men – and later, to a much lesser extent, the women – who embarked on the perilous North Sea oil adventure, represented a polyglot of nations, and still do. They were led by old hands who had learned a tough trade the hard way in the United States, South America, the Persian Gulf, the USSR, Africa, the Netherlands and more recently in the southern gas fields. Across this most cosmopolitan of industries there were Americans, Dutch, French, Italians, Scandinavians, Canadians, Australians, South Africans, Spaniards and Moroccans. There was also the indigenous workforce, the British who came from the south upwards, Norfolk, Lincolnshire, Nottinghamshire, the Midlands, Yorkshire, the north of England, from Wales and from Northern Ireland, and once the caravanserai had crossed the marine extension of the border, from all parts of Scotland. For all of them, exploration offshore was a learning process, but for none more so than the raw recruits who were either undertaking their first employment or leaving behind secure

THE OILMEN

Plate 8.
The inseparables, the young farm boy Rob 'Swede' Lingard and his rig floor mate, 'Little' Joe Dobbs, former hotel wine waiter. *(Grimsby Evening Telegraph)*

occupations and trades. Few knew much of this new industry or about its potential. Many thrived on the work and survived to see oil take its place as one of the major contributors of great wealth to their country's economy. Others dropped out and never returned. The veterans who first ventured out of Aberdeen, the northernmost outpost of the industry, all have different versions of their introduction to this strange and exciting industry. Swede Lingard was one of those who prospered.

'I had been working on the farm since I was twelve, as you do. Started in the North Sea with the Santa Fe Drilling Company. Me dad was all for it. He said he didn't think I would stick it that long, y'know. "Well," he said, "go and have a go," because he had to sign all me papers, because the age of

Plate 9.

Offshore, one of the early helicopters, the Bristow Wessex 60, is given a routine maintenance at Tetney Lock. (*Bristow Helicopters*)

consent was twenty-one then and I was only seventeen and a bit. Anyway we worked on Platform A in the West Sole field. That was in the heyday, you know, everything was like pioneering. It was absolutely brilliant in those days. The thing was that if you could do the job, you got the job. Most people started as roustabouts – and then you worked your way up. Then Santa Fe shut down and took all the other people abroad with them to Libya. I couldn't go because I was under the age of consent, y'see. Some of the others didn't want to go, so we were taken to Tetney heliport (outside Cleethorpes in Lincolnshire) by the Santa Fe drill superintendent, to the BP operations manager. He told him, "I have three, four good lads here – do you want them?" The BP man he said, "We'll take every one of them." So we came off Santa Fe's payroll one week and on to BP's the next week. That was on to the three-legged thing – *Sea Quest*.' Swede, who got his name because of his agricultural background – offshore people are partial to nicknames – forged a lasting friendship with that crew, particularly with Joe Dobbs and the late Bill Drewery. 'Joe were little, about five foot nothing, but by, he was built like a brick outhouse.' After working in the southern sector they followed the *Sea Quest* north.

BP were based in Dundee at this time but the crews left from Aberdeen, where the company had a small office above a Wimpy bar on Bridge Street in the city centre. The oilmen flew from the small wartime airfield on the outskirts of Aberdeen at Dyce, where the helicopter company Bristow, who had begun ferrying men and supplies out to the southern gas fields, had set up an operation. British European Airways Helicopters (later British International Helicopters) had established a base earlier. 'A Dormobile van used to pick us up and take us to the heliport. That was down a concrete road track next to a farm and down to the old aerodrome. They were still using Bristow's Wessex Whirlwinds, floats on and canvas seats. You got in and put on a Mae West, one of them lifejackets. They used to come in and say, "Don't nobody trip the bluidy door off." Old Bob Balls used to be the pilot – he actually won a medal for bravery. Old Bob used to give you a bit of a safety drill, do this and don't do that – all basic stuff. We must have flown thousands of miles with him. Anyways, we had a bloke who, when we were flying along, used to cock the door open and have a wee. Old George woke up one time in a bit of a daze because everybody had been out the night before for a few beers. In them days the door handle and the ejection handle was at the same place. So when he opened the door, it went swoosh like a whirly top and into the water. Old Bob Balls, he was shouting, "What the hell fell off?" And we shouted it was the door. So he said, "Ah well, we're only ten minutes from the rig so we'll just keep going."' More than three decades on, the pioneering helicopter pilot, Bob Balls, who started his civilian career in the Middle East, has only vague memories of that door episode. 'I lost two doors. One was while I was in training in Culdrose, in 1959, but it wasn't actually a door, it was the window. That fell in to some woods. I think – yeah, I think I lost a Wessex door. All you can do is carry on. You are not going to stop for the sake of a door.'

Little Joe had taken a different route offshore. He worked as a wine waiter in a Cleethorpes hotel, a favourite watering hole for the oilmen based at nearby Tetney Lock. Their tales intrigued him and he applied to Santa Fe and was hired. 'I left my job on the Thursday and went out on the Good Friday, 1966. There was only one rig and platform out there at the time. That was for gas in the West Sole field, which had been discovered by the *Sea Gem*. We thought it quite something that we got gas and there was talk obviously of oil being there and we knew it was going to be explored. We eventually moved up to Aberdeen.' Swede said that everybody thought they were only going to be in Aberdeen a couple of years before moving on.

A young trainee assistant driller from Clydeside was part of the rig crew on Shell's *Staflo* from the beginning. Sandy Clow, who was twenty-three, had

Plate 10.
The early European heliport at Dyce Airport, Aberdeen, in the 1970s and the arrival of the Sikorsky 61Ns which took over the North Sea transport role.
(Malcolm Pendrill Ltd)

served his time as a welder and in 1963 worked for John Brown, who were building rig jack-up-hulls for American companies. 'Then I read that a company was looking for welders, it was Shell. They sent me to Holland to train and then I was sent to Aberdeen.' Shell had a small base for their seismic surveys early in the 1960s. The main thrust of their operation still came from England, at Lowestoft and Hartlepool. 'There was no drilling rig at that time. It was being built in Middlesbrough – this was the *Staflo*. Going offshore from Aberdeen for the first time, that was a real experience. Where the main airport is now used to be the heliport, and on the other side was the main airport building, which was wooden. One half of the Nissen hut we were in was the university flying club and the other half was British Airways Helicopters. With one helicopter. They would pull the thing out, an S61, and get it started. Actually, sometimes we helped to pull it out ourselves. BA

Plate 11.

The second of the North Sea drilling 'queens', *Staflo,* in an aerial view while operating for Shell España in September 1976, offshore San Carlos de la Rapita. *(Shell International Ltd)*

didn't have fancy stuff for weather protection and that caused a lot of delays, especially in the winter. So sometimes we had to crew change by boat, which was all right for the ex-fishermen, but for others it was horrendous. It was really rough out there. You had to go up on to the rig in baskets called "Billy Pughs". More often than not you got soaked.'

Ordeal by basket over angry seas was how Keith Johnson, who was running a diving operation in Aberdeen, remembers going offshore. 'Once, I was heading for *Sea Quest* and we mustered in the rail station car park – the BP office was just up the street. Old Bob Dyer was the manager. We couldn't go by helicopter as it was fogged in. So, he says, "Right, you are going to go out by boat." There must have been about thirty of us. These guys said, "We're not going out by boat." So Bob, he says, "The people who are not

Plate 12.
The infamous Billy Pugh – the basket system of lifting oilmen from supply boat to rig – precarious when a heavy swell was running.

going out by boat, stand over there. The rest of you over there." Well, us divers and contractors, we knew we had to go because otherwise we would have been out of work. The other guys were all fifteen years in BP tankers or some such. "Right," he says, "you lot are fired." One of them said, "I have been fifteen years with BP." Bob says, "No, you were fifteen years, now you're fired – we are going by boat." So they all started to move over. Well, we went out by supply boat and they took us up by that old net. Trouble was, if you had upset the crane driver at any time, he would bang you against the side.'

Young Mike Waller from the Buchan village of Inverallochy, just outside Fraserburgh, saw in the industry an opportunity to make more money than in the traditional pursuits of the area. 'My whole background was in

THE OILMEN

farming, although I went to the fishing for a while.' He eventually got an interview with Shell. 'But the personnel guy told me he was sorry but I was too young because of the insurance. I was just leaving and Big Jock Munro – from Nairn – the head of the operation at that time came in. The guy explained about the insurance. Jock asked me what I had been doing. I said I had been brought up on a farm and worked at the fishing. He said, "You are a big, strong-looking loon." So that was me started. First I was kitted out in the gear at Cosalts in Market Street, Aberdeen. Then it came time to join the rig. I remember it was a Friday. I arrived at the Shell desk at Dyce. The *Staflo* was up in Shetland, so we went up in Air Anglia to Sumburgh. It was still small then, in fact the sheep were still on the airfield. We boarded the helicopter and I remember flying in thinking, "God almighty, he will never land this bloody thing on there." But we landed, and it was a beautiful day.'

Laurie Horobin, BP's senior operational geologist in Yarmouth, spent a lot of time on *Sea Quest*. After graduating in geology, he worked in Africa before arriving in Aberdeen in 1972. 'I moved sideways to the drilling department as senior planning engineer and went offshore. *Sea Quest* was good, you never hear a bad word about her. [She was] very much modified for UK conditions. And the crews were very, very good too, most of them had known each other for donkey's years from the Middle East. A lot of Fifers, most of them had been working in the coalmines and it was a time of pit closures.'

In the winter of 1967–68, BP and Shell started drilling and the weather was a revelation. The southern area had been a mere rehearsal for the central North Sea and east of Shetland, where the currents of the Pentland Firth, agitated by contact with the Atlantic Ocean, clashed with the North Sea, creating a climatic zone totally exposed to the icy blasts from the Arctic. The North Sea is in reality an arm of the Atlantic, but with hazards of its own making. More than half a million square kilometres, the largest section is between Scotland and Scandinavia, encompassing the Central, Moray Firth and Northern oil sectors. An all too brief summer period relents in what oilmen call 'weather windows', when temperatures range from 13 to 18 degrees Centigrade. In the winter, with values reaching only 4 degrees, climate and wave power are equally unpredictable. Offshore workers had heard of the hundred-year storm but thought it was merely a tale. They quickly discovered such storms are all too real and dangerous. Again the fishermen could have told them that.

'In the North Sea, the big storms come every year, not every hundred years and you get one every six months.' That was Yorkshireman Ted Roberts' experience when he went offshore as installation manager during

Plate 13.
The *BP Forties* – the lifeboat gifted to the RNLI by BP – hits heavy weather. *(BP plc)*

the construction phase of the first Forties platform. 'You get a hundred-plus mile-an-hour wind and 90-foot seas. So the structures have to withstand them and the men have to operate in that kind of weather. Nobody in the world had ever worked in waters of that depth before nor in that kind of hostile environment.'

Michel Euillet was on a platform in 104 miles-per-hour winds. 'The weather was more than a shock. Not the cold and not the snow, but the actual sea state.' This was the unexpected giant wave fishermen describe as a 'lump of water', a confluence of waves rising in a towering column of water. 'It was in fact two or three waves piling on to each other until in the end it was one wave. I was told it only happened every two or three years, but I saw my first one in 1973. It was extremely unpleasant because you could watch it coming. I was on a SEDCO rig up where Brent was discovered. We saw something like a mountain of water coming very, very slowly and we

THE OILMEN

didn't think we were going to ride it because the semi-submersible was attached to the seabed. Well, we did ride it, but that particular wave hit the spider deck, which was probably close to 80 or 90 feet above the water line – and wiped out whatever was on the deck, a totally open structure. The blowout preventer (BOP) was not down, fortunately, and we actually had no damage, but the creaking of the whole platform was rather spectacular, with the deck twisting. The strange thing was that the weather wasn't that bad. There wasn't even a strong wind.'

Despite their great size compared to the fishing vessels, by their static nature rigs and platforms are more vulnerable. Just how vulnerable, Canadian subsea engineer Bob Archison discovered somewhere on block 210 – destined to be the Brent field – when his rig was struck by a violent storm. The rig was SEDCO 135G, which had travelled from the sub-tropical climes off Darwin in Australia to the near-Arctic conditions of the North Sea. 'As a Canadian I am used to lots of cold weather. But there were many times in the North Sea when you wished you were somewhere else. In late 1972, we were moving to a new location and landed in a hell of a storm. We were being towed by a big Smit Lloyd tug when the towline broke and we began to drift. The tug stayed with us but the storm was so bad there was no hope of getting connected up again until it died down. We drifted in a north-westerly direction towards Iceland for about three days. We were getting further out of helicopter range from Sumburgh. The skipper on the *Lloydsman* told us on the radio he was going to stay with us. But there was a problem. "I'll stay until we get to Icelandic waters. But, do you remember the Icelandic Cod Wars in the late 1960s? During that time I rammed a couple of their fisheries vessels and there was a price put on my head, so I can't go into Icelandic waters." As luck would have it he didn't have to. A day or so later we got connected up again and everything was all right.'

Almost anyone who has worked offshore has suffered a bad weather experience. Painter Graeme Paterson freely admits there have been times when he has been scared by the ferocity of the weather. 'I know guys who have never come back offshore. It is absolutely horrendous. You can't go outside. You can't open the door against the wind. They lock you in the accommodation quarters and won't let you out. The problem is there is nothing to stop it.' David Robertson was a rig welder who sometimes spent 38 hours out on deck at a stretch. 'But you couldn't work in the worst weather and conditions. The wind is the thing and when you are pulling stuff against an 80-mile-an-hour gale, everything is twice as hard. And if you are cold and wet you are bound to be a lot more tired. I was the rig welder and I had to go to the top of the derrick or down one of the legs, hauling burning

Plate 14.
A typical sea state that the offshore workforce have come to dread, making seaborne rescue attempts almost impossible to achieve. *(BP plc)*

gear and welding cables. I couldn't do it now – it was so physical and it is still the same. A machine can't go up there and do that.'

In the far north-east of Shetland, near the Heather field, is an area known to local fishermen as the Devil's Bowl because of its fearsome reputation. Ken MacDonald, who spent several years as a general superintendent offshore, tells of one ten- to twelve-day spell when wind speeds hit over 100 knots, gusting to 120. 'When you consider a Force 12 is 64 knots, you will understand what was happening. The actual wind direction went through 360 degrees and the safety vessels had to abandon their watch and head for Lerwick.' The sheer awesome power of nature was particularly frightening on the fleet of various oil-related ships. Barry Matthews was on one of the bigger survey vessels ordered to stay out because there was a weather window coming. 'What they didn't tell us was that a Force 11 was coming before that. Now, this was a big boat, with two pontoons and a deck well up in the water. But the waves were smashing into the bottom. You could feel them as we headed into this Force 11. Engines full safety ahead – and the vessel was slowly drifting backwards. It was unreal.'

Another fallacy about the weather offshore concerns the visibility. Helicopter pilot Bob Balls described typical flying conditions: 'It depends what you mean by bad. You have high winds and you have poor visibility – either or. And on occasions you can have both wind and fog. Everybody who's flown in the North Sea will tell you that. The weathermen say you can't. But you can. We have all seen it.'

There is another potential hazard offshore. Oilmen apprehensive about helicopters ditching or about falling overboard from ship or installation live in dread of hypothermia, which can overcome a victim even after brief immersion in the bitterly freezing waters. 'The North Sea is as good or as bad as you want it to be.' Joe Cross has been recognised as the world's foremost expert in survival and safety training for the offshore oil industry. 'It is a very difficult survival environment. You can simply fall in the water and for no other reason than the temperature and waves and wind you are in serious trouble.' Sandy Clow framed the consequences in starker terms. 'There was no survival. You went to the heliport in just your normal clothes, and got on the helicopter. The life preservers were under the seats and the analogy was that it was the same as getting on a plane to London. If you went into the sea there was very little life expectation, because you wouldn't last long. But that is just the way it was.'

This, then, was the unpredictable reception that awaited the crews of those early drilling rigs, initially just *Sea Quest*, then *Staflo* and the *Ocean Viking*, when they headed out in 1967 at the start of the northern exploration campaign. In the first round Shell looked at blocks in the central area, but not in the acreage they had licensed further north. It was being surveyed, but it was relatively unfamiliar.

The Second Licensing Round in 1965 had swiftly followed a General Election won by Harold Wilson's Labour Government. A priority for the new Cabinet was to attack the perennial balance of payments burden; oil imports alone contributed to an annual deficit of £300 million, so the Government were anxiously seeking an indigenous supply of the fuel. Their intent was to encourage the exploration companies. Another 1,120 larger blocks were therefore brought into play, but there had been fewer takers, with forty-four companies chasing 127 areas. Shell leased the biggest share and they drilled indeterminate wells on two central blocks. BP gave no indication of drilling in their section. There was still widespread uncertainty about what actually lay beneath the seabed. Although the necessary Palaeocene sediments, conventionally prime sources of oil, had been identified, intensive advance geological and seismological surveys had indicated little sign of hydrocarbons. Not until more holes were drilled was the promising sedimentary evidence uncovered. Even then, geologists would argue for another decade about the actual formations. The North Sea kept its secrets well, frequently defying accepted textbook analyses. To their credit, geologists and geophysicists rapidly developed new techniques to accelerate the survey process.

Gradually, a conveyor belt of supply and support began to reach out into the North Sea from Aberdeen and Lerwick as a frenzy of activity built up offshore. The first seismological boats, contracted to Shell, began working out of Aberdeen from 1962, followed by increasing numbers of service boats. But the North Sea had begun to take its toll. In November 1969, Liberal MP David Steel was told by Roy Mason, the Minister for Fuel and Power, that the worst year for accidents and fatalities since exploration had begun had been 1965, when thirteen had been killed (the loss of the gas rig *Sea Gem*), with three deaths in 1968 and one in 1969. Setting a pattern for future operations by the national search and rescue services on land as well as at sea, a helicopter lifted an injured Sunderland man from a rig in the North Sea. That same winter, aircraft tracked down an Aberdeen-based Dutch oil supply ship, blown off course on her way to *Sea Quest*. Five days later two Italian divers were flown ashore from *Staflo*, suffering from decompression sickness. The unique form of North Sea transport was beginning to clamour noisily for space at Dyce. In January 1969, British European Airways Helicopters 'made the furthest journeys so far in any of the North Sea operations'. From December 1967 to November 1968, the BEA Sikorsky S61N unit carried out 400 flights to *Staflo*, 110 miles from Aberdeen.

Staflo's crews had spent two years drilling one frustrating dry hole after another. Miles Bowen, exploration manager in London in 1969, said in *Tales from Early Exploration* that the general view was gloomy and pessimistic and that exploration was doomed to failure. By then, nine fruitless holes had been dug. Dick Parker, a Shell executive, later recalled, 'At the end of the day we drilled a golf course off Aberdeen. It was a hell of a learning curve.' Shell's London-based exploration and development department couldn't secure additional budget funds and there was a bruising battle in the boardroom to continue drilling. Fortunately for the future of the company, the board decided to allow *Staflo* to plough on.

Assistant driller Sandy Clow was not impressed with his first experience of *Staflo*. He categorised it as a basic floating platform with a land rig stuck on it, like the vessels that drilled in the Gulf of Mexico. 'It came out of the shipyards on 17 December 1967, and went straight out to the North Sea. We started drilling, and it was really rough. Operating it was just horrendous.' Shell had brought people from America to advise and help. 'The bottom line was: they went out, stayed a few days and then said, "Well, we have never seen anything like this and we can't tell you anything. You are on your own."' In 1968 operations improved. 'Things got better as people got used to the rig. We just had to make the best of it. We were all new to the North Sea. A lot of the crew had been in Nigeria, where they worked on swamp barges. They hadn't a clue what the weather was like. About 60 per cent were Dutch, and only 30 British.' Sandy left after two years. 'We had shows, but nothing considered as a discovery. Then, just before I went, you started to see cores coming out with oil dripping from them and they would take wireline samples to see what there was. Characteristically, it was all kept secret. The finds weren't actually significant, but I can still remember the floor being cleared while they broke out those cores. Then it was all coded messages and the cores were sent to London. We were kept in the dark. It was the mushroom syndrome.' That syndrome was endemic in the industry, where all kinds of ruses were used to conceal drilling results. BP engineer Jim Jenner was flummoxed by one unique code. 'The first hint there was anything significant came in a radio conversation between the operations manager onshore and the toolpusher. But it was conducted in Farsi, an Arabic language. This was because of the need for secrecy – communication was by a single-side band radio. These guys had worked in Persia and knew enough of the language to use it as a code. That floored me because I hadn't been there.'

Shell had their own system, described by Mike Waller as a little code-breaker machine like the famous wartime Enigma. 'It consisted of a piece of card – with the alphabet and the numbers; the code changed every day. They sent the coded report and the guys would discuss it on the radio according to the numbers. Then, talking about something else, they would say, "Well, we have got to 10,000 feet," and the secrecy broke down there.'

This obsessive reluctance to disseminate information also appeared to extend to the future Conservative Prime Minister, Edward Heath, in 1968, when company chairman, Sir David Barran flew him out to *Staflo*. *The Press and Journal* industrial editor, Ted Strachan, was among the journalists invited. The exciting new industry did not make much of an impact on Ted, however, nor, as he noted later, on the Leader of the Opposition. 'His Government later seemed strangely laggard in response to opportunities

opened up by the actual discoveries of North Sea oil.' A clue may be found in an election speech in 1974. 'If you want to see the acceptable face of capitalism,' Heath said, 'go out to an oil rig in the North Sea.'

Backbenchers continued to harry ministers for details about the growing industry. In July 1969, Hector Hughes, Labour MP for North Aberdeen, received a reaction from the minister which revealed a curious lack of knowledge. 'I do not think that a White Paper is necessary,' he said. 'I was pleased to hear a Scottish voice [sic], because Scotland will now be involved. BP and the Gas Council have applied for licences for the west coast of Scotland – the coastal area called the Minch. I hope that Scotland will benefit from that.' Only the day previously, the same minister was forced to deny accusations that he had described in a speech the millions spent on exploration as 'abortive' and 'wasted'. He added, 'There is a very high risk in the North Sea, which is a very hazardous area for drilling operations. Consequently, a lot of dry holes can be drilled without any return. Money can be wasted in that sense.' According to the hapless Mr Mason, fourteen companies had ploughed in more than £25 million and *they have had no return at all and have found no gas*' [emphasis added].

Two months later, the oil companies proved him wrong. *Sea Quest* was contracted to drill in a block in the central area by the American company, Amoco. Jim Jenner, a senior drilling engineer with BP, says, 'Oddly enough we seemed to have run out of locations and BP really didn't know if there was oil in the north.' So, for a management reluctant to commit themselves to their own blocks, it was ironic that in September, the rig which Swede Lingard calls 'the three-legged thing' found oil in commercial quantities – for another company. The field called Arbroath was the very first in the Northern British sector. But they had been beaten to the finishing line by the first find in the North Sea as a whole, by an American wildcat drilling company InterDrill (Contractors). On 23 June, they struck oil across the meridian line – the small Norwegian Cod field – unassailable proof there was oil in commercial quantities in the northern sector. The Amoco field's place in history was never revealed at that time. It was four years before the well was appraised and it took twenty-one years for the field to be considered financially viable and come on stream. An adjacent oil-bearing structure proved in 1971, was christened Montrose, and it was first to come onstream in 1976. BP later acquired the two by virtue of buying Amoco. The cluster of wells, including Arkiongut now owned by Paladin, is known as MonArb. Little Joe was on *Sea Quest* when the Amoco field was found. 'We weren't allowed off, but in fact, it coincided with some very bad weather, so we couldn't get off anyway. Then we had a bit more excitement. A boat was

[39]

THE OILMEN

Figure 2.
The North Sea in the early 1970s – the first tentative discoveries that launched the UK oil and gas industry. *(Greybardesign)*

sinking within our rig space and two air–sea rescue helicopters were sent from Denmark, with RAF Shackletons organising the rescue. They had to bring the crew on to our rig. Fortunately, there were two helicopters, because one developed problems and landed with minutes to spare or it would have ditched. This was right in the middle of the big oil find.'

But the most significant find, and that which was generally accepted as the first major commercial discovery, was made by the semi-submersible *Ocean Viking*, working for a multi-national consortium of American, Norwegian, French and Italian companies, headed by Phillips Petroleum. At the end of November, they discovered what was to be a massive oil and gas field hard on the meridian line in the Norwegian sector, 130 miles from the coast. This was Ekofisk, a bounty of 2.8 billion barrels of reserves in the North Sea that set in train the transformation of the Scandinavian country into a wealthy oil province. At last the companies began to believe in the potential beneath the Northern seas and world attention turned to the UK. Encouraged, BP moved *Sea Quest* to the area that was to be Forties, close to

Plate 15.
Ekofisk, the giant find by a consortium headed by Phillips – the last gasp attempt after a discouraging series of dry holes. The field laid the foundations for the hugely successful Norwegian oil and gas industry.
(Conoco Phillips)

where Shell had dug so many dry holes.

The finding of Ekofisk was a story of sheer determination. Phillips had drilled thirty-two holes without success in an operation that swallowed money at a frightening rate. They were becoming desperate. Diver Keith Johnson said one of his Comex team was on board the *Ocean Viking* as it battled through terrible weather, nearly capsizing at one point. 'Phillips were going to pack up and go home, but as there was contract time left, they decided to drill just one more well. And that was Ekofisk. Incredible.' Ted Strachan broke the exclusive story in a few paragraphs on the front page. 'It actually shared attention with the closure of Aberdeen's famous Rubislaw quarry. The end of one traditional industry, granite, and the beginning of a brand new one, oil. I picked up the story from the Phillips guy in London, Paul Tucker. He told me what it really meant.' If other parts of the northern seabed concealed the same prolific hydrocarbon-bearing structures, the North Sea would be the next important major oil province as well as a secure source of fuel. Financially shackled Britain and a Scotland in industrial decline had unwittingly fallen heir to a cornucopia of riches, apparently dispensed by the Almighty, according to the politicians, at least. On a visit to Forties, Labour Prime Minister James Callaghan said, 'God is once more on our side.' Sir Andrew Gilchrist, chairman of Highlands and Islands Development Board, was more pragmatic: 'North Sea oil is God's last chance

for the British.' The next find, at the end of 1970, provided complete confirmation. BP had shared information with Shell about two blocks where *Staflo* had been drilling, which also looked promising for neighbouring locations held by BP since 1965. So they moved *Sea Quest* across to where they eventually established Forties, the second of the three North Sea 'elephants' (the third was to be Shell's Brent).

At the beginning of 1971, Shell and *Staflo* finally made the breakthrough in a block further south, one of eleven acquired in the third licensing round. The find was christened Auk, the first of an alphabet of seabirds. Mike Waller was a raw young roustabout on the rig. 'What I remember of that day was the coring. There was obviously something going on – because all us workers were sent off the floor. The core looked like a 90-foot barrel. It was taken by the bosses themselves to the geologists' area for examination. It was all secret. That only lasted for one lot, because they were heavy, but they were still covered up. Then we did a well test. You obviously kent something wis going on. When you had a find like that, the Schlumberger guys came out and ran electronic detection tools to find out whether it was water, oil or gas. When it was good, they always used to hand out cigars.'

Encouraged at long last, Shell moved *Staflo* 150 miles north, up to the 61st parallel to the east of Shetland to what appeared to be a promisingly large prospect. This time security was total. The company refused to announce any findings before the fourth round of licences, due in six weeks. Shell had their sights on an area adjacent to block 211, where they had started drilling. The justification for holding back the news was that they had found the huge Brent field, which later gave its name to the benchmark North Sea crude oil price. With reserves of 2 billion barrels of oil and 4 trillion cubic feet of gas, it was bigger than Forties or Ekofisk. But Shell weren't satisfied. They wanted to mop up whatever else was present, in particular the next block in what would be the central field in the immensely fertile East of Shetland Basin. Mike Waller recalls when his rig struck Brent. 'It was August, beautiful weather, a fantastic summer, long hot days. We were coring there for ages, so we knew there was something on. What confirmed this was when six chopper-loads of Schlumberger tools came out. It was a Saturday, I remember. Well, this thing was logged to death – so everyone eventually knew that was a big one.'

That crucial fourth licensing round in 1971 was orchestrated by the new Conservative Government. Faced with a threatened crisis in world energy, the Tories had promptly reversed their predecessors' policy of maintaining tight control on the oil industry. Their unprecedented ploy for the new round was to stage one of the most unusual events in British economic history, more

Plate 16.
Shell Esso's first discovery in the northern region – the Auk field. Deliveries are being winched on to the field's A platform from a supply vessel. The platform currently produces, meters and pumps oil from an exposed location single buoy mooring system. *(Shell International Ltd)*

characteristic of the fevered excitement of a Sotheby's sale than of the prosaic award of oil and gas licensing. The remaining 278 blocks were released, but the departure from previous practice was to allow fifteen sealed bids for cash. With their privileged knowledge about East of Shetland, Shell were convinced another giant field was possible on a neighbouring location, block 211/21. So this was the target for a sealed offer with their partners Esso, who estimated it was worth £100 million, justifying a bold gamble. The auction, held in August 1971 at the Department of Energy, attracted worldwide interest; in the audience oil magnates of considerable stature, including the multi-millionaire American J. Paul Getty and international speculator Armand Hammer of Occidental. The first bid was £4 million for a block, which became Mobil's Beryl field. Finally, Angus Beckett, the Under Secretary of State for Energy, read out Shell's offer for what was known ever after as the 'Golden Block'. To the astonishment of an audience not unaccustomed to spending colossal sums of money, the bid was £21 million.

Unfortunately, Shell were to be disappointed – the wells on the block, named North Cormorant in 1974, possessed nothing like the 'gold' of Brent.

Plate 17.
Night view on Brent Charlie. *(Shell International Ltd)*

But eventually, the revenues from the combined Cormorant fields – the South well was discovered two years earlier – recouped the company's ambitious outlay. The energy auction angered Opposition politicians. Only £37 million had been raised from the sealed offers and £4 million from the other blocks leased to 212 companies. The Parliamentary Committee of Public Accounts later said that the Energy officials were too lenient with companies already making huge profits from the southern gas fields. The sealed bid gambit was never attempted again. Significantly, the other oil companies, poker players to a man, were displeased that Shell's extravagant bid had revealed the potential value of their own hands. They quietly stopped complaining that the Government's fiscal demands were damaging their ability to develop the burgeoning UK province.

Behind that spectacular fourth round is a revealing anecdotal tale involving Armand Hammer and the British Government. It came from independent oil consultant Alex Barnard, whose family has a long history in the industry, and who graduated from Skerries Technical College in Glasgow in the 1950s. He worked first for Texaco, later for P&O's oil division and Global Marine. 'An island offshore from Abu Dhabi was disputed territory with Iran. Occidental asked the UK Foreign Office if it was safe to drill there despite the dispute, and Hammer claimed they were given the go-ahead. So they began drilling. Then an Iranian gunboat ordered them off. The drilling superintendent said they had permission, but with immaculate aim, a shot from the gunboat severed the starboard anchor chain. The crew made the well safe and got out. A furious Hammer took out a lawsuit against the British Government because of the advice from the Foreign Office. The minister, Sir Alec Douglas Home, sent an invitation to Hammer, who went with his vice-president. Sir Alec said, "I believe you are interested in the North Sea?" As Hammer was leaving he was handed a piece of paper and told, "You know the golden blocks, apply." They were Piper and Claymore and they were huge. Hammer withdrew his lawsuit.'

Throughout the next decade, the cost of drilling operations in the sectors rose to £1.7 billion by 1977, and continued to rise into the 1980s. The discoveries were coming thick and fast and were matched by the pace of development, which added a further £5.6 billion to the total bill for the North Sea. By the middle of 1970, the first UK oil finally came ashore, surprisingly not from a major producer but from an independent.

[3] Life Offshore

Hard graft and long hours

The monstrous steel island rising up from the seabed in the northern North Sea was to be workplace and home for a Dundee plumber for the following sixteen days and, on and off, on installations like it for the next seventeen years. Jake Molloy's immediate impression as he landed on the helideck of the Ninian North Platform was unforgettable. 'I can still see it. It was just awesome – that is the only word I could use.'

Far less inspiring was the first encounter with an oilrig for young Roy Wilson from Aberdeen, who had just joined the ARCO drilling company. 'There was a rig called the *Glomar V*, in Aberdeen harbour, a right bucket, and the ARCO people said, "Look, we'll give you an opportunity to go on a drilling rig and get some idea what it is all about." I saw this great rusting ship which was marginally better inside, but for the most part, it was pretty seriously depressing, and I thought, "My God, what kind of a business have I got myself into?" It was entirely the wrong foot to start out on – but it didn't dissuade me from carrying on.'

As the 1970s rolled out, *Staflo* and *Sea Quest* had been joined by dozens of other explorers in the North Sea. Remarkably, as late as January 1973, politicians were still being urged to be cautious. The Department of Trade and Industry said in a report that the importance of the discoveries [Ekofisk and Forties] remained to be established. 'After twenty-one of twenty-five wells drilled in the British sector had been found to be dry (by the end of 1970) . . . there was no certainty that a major oil bearing area existed.' But the companies were no longer in doubt. By 1973, there were eleven rigs in action; a year later that number had trebled to thirty-nine rigs working, thirteen of them new semi-submersibles and one a dynamically positioned drill ship. A host of ambitious players – some new and others from the southern sector – had arrived on the scene. Texaco, Chevron, Occidental, Mobil, Total, Conoco, Phillips, Elf and Marathon were all joining the race. Among the smaller newcomers was a company called Britoil, operating from Glasgow.

Plate 18.
The North platform's topside heads out towards Chevron's Ninian Field.
(CNR International)

THE OILMEN

In 1973, the industry, desperate to establish a secure strategic source of energy in the West, received another incentive; the Yom Kippur war between Israel and her Arab enemies, once again paralysing the international fuel market. To break the United States alliance with Israel, the Arab-dominated Organisation of Petroleum Exporting Countries (OPEC), formed by the oil-rich Gulf states in 1967, deployed the oil embargo weapon, cutting production, raising prices and staunching the flow of oil to the dependent West. Not for the last time, the dynamics of Middle East oil politics would influence capital investment in the North Sea. Because of the generous terms of the original offshore concessions, when OPEC imposed an oil price five times higher the companies stood to make a great deal of money.

Harold Wilson's Labour Government of 1974 were determined to dismantle the previous *laissez faire* policy of the Tories. After turbulent negotiations with the industry, two new regulatory bills to squeeze out more revenues became law in 1975; the Petroleum Revenue Tax Act and the Offshore Petroleum and Pipelines Act. New Energy Minister, the ultra-left wing MP, Tony Benn, introduced the latter. The tax take was not as penally restrictive as had been expected. A single rate of 45 per cent of gross production profits before 52 per cent corporation tax allowed the Government 70 per cent of a field's earnings. The next political step was the formation in 1976 of the British National Oil Corporation, which took a stake with participatory holdings in fields, including Hutton, Thistle, Brae, and Dunlin. BNOC could also buy 51 per cent of all oil and gas production. In a device to control prices, the companies could buy it back at no extra cost. The Glasgow-based corporation, which employed a thousand people, lasted nine years until the Thatcher Government closed it down. The private independent Britoil took over its oilfield assets and in turn was bought over by BP.

Onshore, BP, emboldened by Forties, had immediately set in train negotiations for land in Scotland; first for the subsea oil pipeline landfall and then for the transit corridor to their refinery at Grangemouth on the Firth of Forth. The construction of the production platforms was also well in hand, two at Laing Offshore, on Teeside and two at a new yard at Highland Fabricators at Nigg, in Easter Ross.

The build-up continued offshore, and in the fifteen years to 1985 the workforce grew from 9,000 to 31,300 before dropping during a dramatic slump in the oil price. But the companies were finding it hard to recruit men with the necessary skills. The first installation manager on Forties, Ted Roberts, interviewed hundreds of people. 'The local authorities weren't just interested in the money but in jobs for the North-east of Scotland. It had to

be explained there would be jobs, but for people who were qualified. To be honest, there were only a few qualified.' Roberts said the same applied to applicants from the north of England, the Midlands and Wales. BP had to instigate crash-training courses, while a Shell Expro and Grampian Regional Council trainee technician scheme produced hundreds of graduates at a rate of fifty a year at its peak. In a 2001 study paper, 'North Sea Oil and the Aberdeen Economy in Retrospect',* international petroleum economist Professor Alex Kemp, of the University of Aberdeen, noted that 'not all [oil-related] employment accrued to residents in the North-east . . . a significant proportion came from overseas due to the lack of appropriate skills within the indigenous [British] population.' In fact, by 1983, 80 per cent were UK nationals, a figure increasing to 90 per cent by 1985.

John Selbie had to bide his time to go offshore. A local man from Dunecht, he was a time-served joiner. His first experience was on contract to Shell to build their expanding offices in the late 1960s. He actually went offshore to convert cabins on Shell and BP rigs. The first trip was to *SEDCO 735F* off the coast of Norway. He and his fellow crewmen went out by supply boat. 'When we got there alongside this huge rig, I'm thinking, "How on earth will we get up there?" Then I see this basket coming down with the crane. The guys just threw their bags in the middle and stood inside the basket. Because of the weather I was there ten days, although my work only took two and half. I couldn't believe it – who was paying my wages for this?' It was two and half years before John secured a job as a roustabout with Shell.

Making a start was far easier for Jake Molloy, a time-served self-employed plumber in Dundee. At the end of the 1970s, he was fitting a bathroom suite for a man who worked for Chevron. 'He told me they needed plumbers offshore to do all the domestic stuff, and gave me a number to call. That was a Friday, and I had to be in Aberdeen on Monday. This tiny wee office was upstairs on Union Street. A girl behind the desk said, "What size chest, what size boots? Grab a hard hat." There was a taxi to take me to the airport, but I had no idea where I was going, what I was doing, or anything. I had come with an overnight bag. I went back down the stairs, kissed my wife goodbye and jumped into the taxi. Out at the airport, a lad told me we were going to Shetland. I had no idea where that was and I had never flown before. We went on to this old Loganair fixed-wing plane and flew up to Unst. It was like a different planet. Then it was into these portacabins and I

---

* Kemp, A.G. and Smith, F., 'North Sea Oil and the Aberdeen Economy in Retrospect', in Starkey, D.J. and Hahn-Pedersen, H., eds., *Concentration and Dependency: The Role of Maritime Activities in the North Sea Communities*, Esberg, Denmark, Fisheries and Maritime Museum (Esberg, 2001).

THE OILMEN

watched what everyone else was doing. I picked up what I now know to be a survival suit, pulled that on top of my gear and marched out to this helicopter, which was pretty scary. No survival drill and we were all new guys. We got this wee yellow bag with a cord to tie round your waist – this I discovered was the lifejacket. It was a SK61 helicopter – no Tannoy on it – and out we went to the platform, Ninian Central.'

There was a longer wait for David Robertson. He is a boilermaker to trade and began work welding and pipe fitting in Motherwell before moving to England in the 1970s. He and his wife actually swapped houses with an English couple living in Aberdeen. He was eventually hired as a welder by Santa Fe on the rig *Bluewater 3*, drilling on the Thistle field. 'That was the first time I was ever in an aeroplane. We were in the Amatola Hotel for three days, because it was foggy. Then we flew up to Shetland in a DC3.' As with Jake, there was no briefing, but the men were in ordinary clothes, no survival gear. 'If you wanted you could take a bottle of whisky; there were no searches. Then it was on to a helicopter and out to the field. I didn't realise rigs floated – I was standing in the galley and thought, this thing is moving.'

The rigs and platforms offshore are like no other form of work locations – steel or concrete archipelagos marooned in a featureless, often hostile ocean, while the mobile drilling vessels represent another alien world for the raw landsmen. There are three types of rigs – the semi-submersible, the drill ship and the jack-up. The last is for shallow water, while the drill ship has mobility but is less stable. Some ships now have computer-controlled dynamic positioning systems, so they can drill in almost unlimited depths. The favoured early semi-submersibles like *Staflo* operated in most conditions to depths of 600 feet. Modern semis are self-propelling and are equally comfortable as either floating production units or drilling rigs. Drilling is a prohibitively expensive exercise, which explains the incessant pressure on the drillers. During the early years it took around £50,000 a day to keep a semi in operation; a well could cost about £3 million and still be dry.

Techniques in drilling are constantly changing but the fundamental elements are: a derrick, a 230-foot crane straddling the well hole and used for handling and stacking drilling equipment; a rotary drill system and drilling mud. Wire ropes run from the crown block at the top to the travelling block attached to a square or hexagonal steel tube called a kelly which connects to the rotary table. The drill pipe and bit is screwed on to the kelly and lowered down the hole. More pipes, the drilling string, are added as the drill bites. The bit is changed regularly and the total manoeuvre of pulling up the string, replacing the bit and resuming drilling is called a round trip, which can take up to 10 hours, depending on depths. A special drilling

Plate 19.
Top left. The essential features of an offshore rotary drilling rig. *(Greybardesign)*

Plate 20.
Top right. The sheer hard physical labour in the early days. *(BP plc)*

Plate 21.
Below. The less demanding work on the drill floor of a modern rig. *(James Fitzpatrick)*

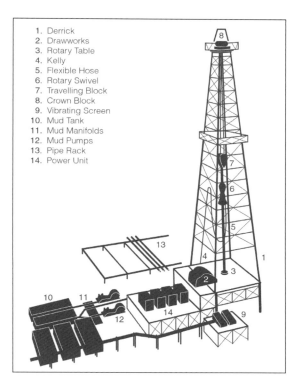

1. Derrick
2. Drawworks
3. Rotary Table
4. Kelly
5. Flexible Hose
6. Rotary Swivel
7. Travelling Block
8. Crown Block
9. Vibrating Screen
10. Mud Tank
11. Mud Manifolds
12. Mud Pumps
13. Pipe Rack
14. Power Unit

THE OILMEN

mud pumped down the hole serves a number of vital functions; to flush out chippings, cool and lubricate the bit, and maintain pressure on the well walls. Logging and analysing the mud can warn of problems in the well. Valves called blowout preventers are set at the head of the well. Thick steel pipe casing is cemented in as lining. Once the oil trap is determined, the first wildcat well is capped with a control called a 'Christmas tree'. Appraisal wells are then drilled, followed by development wells which eventually produce the oil or gas.

Work on the early rigs was dirty, often dangerous and physically demanding. This was why Shell's Jock Munro had selected Mike Waller as 'a big strong loon'. Mike said, 'The *Staflo* crew were all local; some from around Nairn, the Jock Munro connection; two from the Fraserburgh area and a welder from Peterhead. Others were from elsewhere in Scotland and northern England. A steady crew was seventy-five, making up two shifts – plus service providers. The drill crew was six guys – five roustabouts and a crane driver – then there were the usual mud engineers and others.'

Kevin Topham, a derrickman who survived the sinking of the *Sea Gem*, said that in the 1960s newspapermen described drilling as 'the toughest job in the world, after mining'. Kevin recalls, 'To do a bit change at 10,000 feet, you had to pull 10,000 feet of drill pipe out, using power spanners on each joint. Then the derrickman would rack the pipe at the top. The man at the bottom stabbed them down on a wooden platform in rows of single 90-foot lengths. His was the most dangerous job. If he got his toes in the way – and one or two of them did – he was minus toes even with safety boots on. The other dangerous part was using the two power spanners with a man on each spanner. If you caught your fingers, then you lost them. You wouldn't believe the number minus toes or fingers. You had protective clothing, but there were no guards except on the revolving drive chain. It was dirty and it was very noisy, but you got used to it. Twelve-hour shifts, seven to seven at night. You did get the occasional quiet shift waiting for the cement to set round the casing. That could be about 48 hours.' Kevin's contemporary, Swede Lingard, revelled in the sheer hard labour. 'It was all lift it, pull it and shove it and we had to throw the chain to spin the pipe. If you got a good gang and you all worked together, it made the job a lot easier, more enjoyable. I suppose it was dangerous, the things we did to get the job done, but at that age you don't care.'

The drilling techniques became more mechanised into the 1970s, but that didn't make them any easier. 'For us,' said Mike Waller, 'the technical knowledge was pretty basic. I don't think it was deliberate. It probably came from working in a Third World country. That is one of the biggest changes –

Plate 22.

Silhouetted in the North Sea dusk: *Stadrill*, the 12,000 ton self-propelled, semi-submersible drilling platform which replaced the eight-year-old *Staflo*. (Shell International Ltd)

Plate 23.
Where there's muck . . . a drilling crew brave the mud on a round trip.
(Allan Wright)

the amount of local people who have progressed. We used to work seven to seven – got up and had the supper and we were going to run casing. In good casing running tradition, it was pouring rain at night – this was on open pipe decks – and my brand-new gear quickly became muckit. So we ran that casing, which involved cementing in steel pipe to line the well, and I remember thinking, what kind of a bloody job is this? Coming from a farming background, I was used to getting mucky – but I thocht fit a fool orra job this is. I hidna clue what was goin' on. For me it was basically tidying up the entire casing, washing decks. That wis generally your life for a while.'

The role of roustabout had changed little by the time another local man, Martin Reekie, an engineering graduate from Dundee, joined Shell in 1980. 'Joined on the Monday and offshore on the Tuesday. That was before you needed any form of training. Started as a roustabout and worked my way up. It was all brand new to me. My first job was cleaning out pump liners and I didn't have a clue what they were for. I was on a semi-submersible, the *Stadrill*, which was Shell's training rig, off the Cormorant Alpha platform. I learned very quickly that a roustabout was the jack-of-all-trades. There was an awful lot of cleaning, and painting, scraping and wire brushing. A roughneck, on the other hand, worked on the drill floor.'

Plate 24.
The derrickman, perched high on his monkey board, 140 ft above the sea has a highly responsible job on a drilling rig, next in line of command after the assistant driller.
(Allan Wright)

A layman's perspective comes from Ray Craig, who worked on a rig to see what it was like. He had already spent a university summer vacation as a supply boat skipper, but when he went offshore he was a trainee lawyer, dealing with offshore insurance claims. 'So I asked to see for myself and I worked for a week for Atlantic Drilling on a semi-submersible as a roustabout. And I thought, "What the hell am I doing here?" I was a 42-year-old apprentice lawyer and these were eighteen-year-old roustabouts. A team of four and there was no hiding place. Roustabouting was murderously boring, basically labouring in quite unpleasant conditions. I was actually let loose on the drill floor for 20 minutes. Can you imagine that happening now? Within the first day there was a Force 7 gale shrieking through the moon pool – and I am stuck in a bosun's chair right over this drop and they swung me out. And they paid me full rates, which was more than a trainee lawyer got.' Ray later became the second 'oily fiscal', the procurator fiscal specialising in offshore cases.

Literally, the next step up the ladder was derrickman, who perched on the monkey board. Among other things he had to rack the pipes. John Selbie had been promoted in 1975 to floor man (Shell roughneck) and then to derrickman. 'He was the leader of the crew, below the assistant driller – and that was about as far as local guys got then. Shell hired graduates or guys

THE OILMEN

with Higher National Diplomas as assistant drillers and drillers.'

Mike Waller worked as a roustabout for eleven months before moving up to roughneck in 1972, tragically because of the death of a derrickman in a fall. 'Shell wanted to give local guys an opportunity to progress and four lads were chosen as assistant drillers. So they needed somebody to fill in. I had never worked a derrick and you were also in charge of the mud systems and the mud pumps – so that was a steep learning curve. You used to pull the pipes by rope and learn how to use your body.' He was very much aware of the perils involved. 'You were about 85 feet above the floor and a total of 120 to 140 feet above the sea. There was a lot of bad weather up there.'

When young Charlie Anderton finally became a derrickman with Santa Fe Drilling, he was always out to impress. 'One of the things I used to do was to throw heavyweight pipes on the run. When the blocks were running past, you would throw the pipe in when it was moving. The heavy pipe was 50 pounds per foot whereas a drill pipe is about 20. If it missed it would crash across the derrick and in theory you could actually snap the pin. It was highly dangerous.' Charlie made it to assistant driller, 'By then I had sussed the highest-paid job was the directional driller. It was the black art. I remember asking the driller, "What is that boy doing?" He said, "I don't know" and I said, "That's the job for me."' With the directional technique, as many as thirty holes deviated from the vertical can be drilled from a single static platform. Charlie eventually trained with Dowell in Argentina and then went to Africa, which he hated. 'I was very fortunate; when I came back I got a job on Beryl Alpha working a double derrick. That was one directional driller looking after two rigs simultaneously on the same platform. I got something like 4 hours sleep a day, so it was really bloody hard. By this time I was twenty-three years old. I remember this driller was quite aggressive and he asked me, "What the bloody hell do you know?" I said, "Well, not an awful lot." "In that case sit down here and I will teach you the basics." And he did. He was a Dutchman, a lovely man.'

Sandy Clow was employed as an assistant driller on *Staflo*. 'The eventual aim was to become toolpushers or night toolpushers. You trained in Holland a year and then got a posting as an assistant driller. You could work up to driller and take it from there. I was about twenty-two or twenty-three, which wasn't unusual, although I was probably the youngest. When the rig arrived in the North Sea, there was an assistant driller and a few people on the mechanical side, but the crews were basically fishermen, or farmers, all local, with absolutely no idea what it was like – everyone was at the bottom of the learning curve.' Martin Reekie also advanced rapidly – despite still being in his twenties – as the industry hit the boom employment era of the 1980s.

'With so much work, you got fairly quick promotion. The crews were all about the same age. Certainly, the drillers on *Stadrill* were ten or so years older than myself, but they had been around longer. I spent some time as a derrickman and then assistant driller, learning on the job.' By then his Shell colleague Mike Waller was working as a driller and the industry had begun to develop new technology to overcome the obstacles of the challenges of the North Sea. 'We had started doing subsea projects. The first at Brent wasn't that successful, but getting the stuff in place was OK. Then we did a single Cormorant underwater manifold and we became more involved with subsea completions and more production guys involved. It was quite a challenge with these people on your rig.'

Martin Reekie went through the same training system. 'But you progressed on the job, trained by experienced people. They hadn't really been taking on too many college people, so I suppose I was a bit unusual. Now you need first-class honours to be a driller.' Martin had been beaten to the age record more than ten years previously by the ebullient Swede Lingard. 'The toolpusher said, "You will be drilling next time you come off. I'll put your money right," and it was done. You never needed an interview. So at twenty-one I became the youngest driller in the North Sea, on the *Sea Quest*.'

Plate 25.

Crew-change by drop-in on one of the Brent platforms.
(Kenny Thomson/ John Greensmyth, Technip Offshore)

THE OILMEN

Next step for Martin Reekie was as toolpusher on Brent Delta. 'I then went to Dunlin as company representative running the drilling operation. On the platforms you get a bit more involved. You are part of the management team, which is tool pusher, production supervisor, and the offshore installation supervisor who looked after the catering and helicopters.'

Swede went toolpushing for three years on permanent night shift before being promoted to senior toolpusher on day shift. 'I think that was one of the best times, because you were able to put a lot of your theories into practice. The toolpusher was the man with the plan, relatively well paid and with a lot of responsibility. You were on a £65-million rig and a drilling a well that could cost about £12 million at that time. At the end of the day, you were on your own. Before I left, they had started to bring in those OIMs [Offshore Installation Managers].' Another who had progressed to night pusher was Mike Waller. 'Then I moved on to contractor rigs as a company man, and I began to get involved in high-pressure, high-temperature drilling – a completely new world to me again. That was a learning experience. You were the only guy there, and the first time or two, it was quite sobering. But you learned to man manage and think your way through it. I did a number of years on these contractor rigs – then went back to the *Stadrill* as senior toolpusher and OIM.'

By the mid-1970s, the distinctive work pattern of the offshore crewman was well established. First it was more than a week on and less than a week onshore, with a working day of not less than twelve hours, although for some rig staff that was elastic. For recruits from onshore industry or commerce it was a considerable culture shock. The cycle was slightly different on the early Shell rigs – seven days on, seven off. Mike Waller said that it was far better later with two weeks and two or even two and three. 'You would have a really hard week on the rig, and by the time you recovered, you weren't long before you were back. There was a lot of travelling, which is tiring in itself. So I liked it fine when we moved to the two on and two off.' Rig supervisor Ken MacDonald's crews worked a much more strenuous pattern. 'In the early days it was pretty horrendous, starting off with twenty-five days on and five days off. Then it developed into two on two off, but you also had situations in bad weather when supply boats couldn't get out, the laundries would be shut down, the food was cut to the minimum. That didn't help.'

From the American wildcatting days to the North Sea, oil has always been associated with power and almost unimaginable accumulations of wealth – as the volume of offshore investment and cascade of revenues revealed during the UKCS boom years. Between 1965 and 1976, some £5 billion was lavished on development; in 1976 alone that was a quarter of the

Plate 26.
A supply boat feeds the SEDCO 135G drilling rig in the sun-dappled waters of the North Sea – a tranquil scene for once.
(James Fitzpatrick)

country's total industrial investment. In the same year the offshore supply market was worth more than £1 billion, while Britain's troubled balance of payments account benefited by £2 billion. Whether or not the industry's frontline soldiery profited in terms of wages and salaries is, however, more difficult to validate.

The general perception has always been that the oilmen must have been earning generous sums, and some undoubtedly did. But calculating the ratio of wages paid to the hours worked, most made little more than they would have onshore. During two spells, however, the companies were described as 'simply throwing money at projects and at people'. The first was in the 1970s' scramble for discoveries; the second during the frantic build up from installation to first oil. To a brash farm boy in 1965 the sudden wealth was unbelievable. Swede Lingard was paid four pounds ten shillings a week at home. 'Working as a roustabout we got six shillings and sixpence an hour – pretty good money. We worked 12-hour shifts, midnight to midday, and it was all hard graft. Ten days on, five days off, and if you were away two weekends over a Bank Holiday, you couldn't half make some money. The first time, after the stamp and the tax, I cleared forty-seven quid. I had a post office savings account with Mrs Stark in the village shop. Santa Fe gave you a five-pound note to get you on your way home and a cheque. The cheque said "Santa Fe Drilling, Chase Manhattan Bank, New York". Mrs Stark took one look and said, "Ooh, I don't know about all that." It was worth about forty-two quid, bright orange it was, skyscraper in the background. The biggest cheque I had ever seen, size and money. Mrs Stark, she was having nothing to do with it. So I opened a bank account in Cleethorpes. On *Sea Quest* I had gone from roustabout to derrickman, ten and six an hour. The second time I came home, I paid cash for an American ex-army jeep off a scrap yard, £70 out of twenty days' work. That would have taken me about twenty years on the farm.'

Not everyone regarded the pay-out offshore as generous. Shell assistant driller Sandy Clow claims it wasn't any better than in the shipyards. 'There was no wage structure in the UK for the crews; our salaries were tied into the only other oil-related business, which was refining. We were classed as international staff and paid £2 a day offshore allowance, so for seven clear days, you got £14, which was taxed. This was a basic salary. Just before I left, they brought in a shift allowance of £200 a year. But that was when they realised people were leaving. The crews were all right, they were getting overtime. So in 1969, three or four of us transferred overseas, otherwise we would have left. It used to annoy me intensely when I went back home to Helensburgh and people would say, "Oh, you are one of these £100-a-week oilmen." There was no way I was getting anywhere near that: £100 a month yes – but not a week.'

Aberdonian Neil Ferguson was trying to make a living at the end of the 1970s as a professional musician in England. 'I got married when I was playing. My wife became pregnant and the trumpet playing wasn't really paying the way, so I decided to return to Aberdeen to my trade as a joiner. Then Dowel Schlumberger were hiring for the gas gathering pipeline from St Fergus out to Brent, and that's where I started.' This was the Far North Liquid and Associated Gases System (FLAGS). He also went offshore as a supervisor for cementing operations. 'As a joiner time-served, I was taking home just over £100 a week, £120 maybe, with overtime for a nine-hour day, Monday to Friday, and on a Saturday morning. With Schlumberger, I was taking home £200 a week – salary and offshore bonus. You got £6 a day offshore for the first fourteen days in a month, and if you did more than that you got £26 a day. Because I worked on land on FLAGS on a twelve-hour shift, after the first fourteen days I was on £26 every day. So that was good money. As a professional trumpet player I was on £30 a week, and £6 a day when we played. It was good fun, but it didn't actually pay the bills.'

The career change to directional drilling was a highly lucrative for Charlie Anderton. 'By the 1980s, I was twenty-three years old and I was paid $375 a day bonus and a $20,000 salary. A young man making the guts of $90,000 a year, renting a flat for £26 a month, I was kickin' ass and making money.' Charlie later came onshore, still only in his twenties, and set up his own company, Andergauge, to market inventions in downhole technology.

The experience of US geologist Bill Glidden puts a different perspective on the commonly held view that most American oilmen earned a fortune. Bill graduated in geology in Oklahoma in 1950 and worked in the States and overseas before coming to Aberdeen with Corelab in 1967 on contract to Shell. He was on *Staflo* when Brent was discovered. 'One time I worked for 109 days with only two days off. I used to leave at four or five in the morning and get home at midnight. I hardly ever saw my young daughters. I didn't even get overtime. Finally, my wife said, "You either quit your job or I am going back to America. This is no life."' Bill took the hint and went to work on his own, contracted to Union Oil on a rig for $200 a day. Barbara Glidden, a charmingly forthright Southern lady, said, 'International Drilling wanted Bill and told me to pick out a company car. We had an old beat-up station wagon with Corelab. When Bill came back a car was waiting to meet him. He signed the contract and they were wonderful people to work for.' Bill and Barbara are living on Deeside in retirement.

Oil tanker captain Alan Higgins became offshore installation manager for Chevron during a hook-up and found he had to be available for duty 24 hours a day – with no overtime. 'You were lucky if you got 3 or 4 hours a night. Everyone else was in the same boat, but they were on overtime and making a lot of money. I earned, in those days £10,000 a year, the going rate, but a drop of two or three thousand from a ship's captain. Within two years, however, I had overtaken that. But the guy who pushed the tea trolley then earned £2,000 a year more than me because he worked from six until ten on overtime.'

Supervisors with other companies like Mike Marray of Shell encountered a similar dilemma. 'Occasionally if the weather was bad up in Shetland, the aircraft couldn't take off. If you waited for more than 36 hours you got what was called a "golden fortnight". Even back in the early 1980s that meant

THE OILMEN

£600. So the guys were always rubbing their hands about these golden fortnights. But the supervisors *had* to get out and you were put on a supply boat, generally taking 12 to 14 hours overnight. By that time, the fog had cleared and the helicopters would be flying again. Meanwhile, you were lifted from the boat by crane. So you missed the golden fortnight and you had a lousy journey.'

As a plumber onshore Jake Molloy had been making from £80 to £120 a week, all the hours, and with call-out on a Saturday. 'Offshore, it was £4 an hour for a basic 168 hours, plus any overtime. When you hit the beach, somebody met you with £30 in cash and a cheque, but then you had to sign the dole, because you never knew about the next job – that was a contractor's life at the time. So you told the job centre you were looking for employment; a plumber, offshore earning £300 a week. That was just not going to happen. "Unless you have something along those lines then I am not interested." But at that time, it was almost guaranteed you were going back offshore again.'

Former Forties installations manager Jim Souter joined from the Merchant Navy in 1974, starting as a mechanical supervisor with the same pay as in the Navy. 'The only gain was that I was only away for two weeks instead of six months at a time. A year later we got a fairly large increase on the coat tails of one of our main contactors, GEC, 20 per cent or so. BP employees were never the highest paid, but then terms and conditions were as good as any. As management, we still got time off, whereas some of our competitors were dragged in frequently. Later some very senior management – from America – thought we had it too easy and that materialised in general cost savings. Peoples' terms and conditions have now eroded quite a lot.'

Ultimately, in comparing the rewards offshore with those onshore the totally different lifestyles must come into consideration. Ronnie McDonald, as a union official, did the sums. 'The old saying was, "If you canna hack it, jack it." Let's face it, people went offshore for cash. I suppose a lot of the guys who had come from factories found it quite lucrative. I don't want to be too critical, the wage was a good one, and the main attraction was the time off. If you worked out the hours however, in the 84-hour week, you actually did 400 hours a year more than the average industrial worker.'

Among the oil professional classes, however, the salary scales come close to justifying that 'millionaire' tag. A 2004 report by the Hay Group found geoscientists, petroleum and drilling engineers were the best paid in the industry. While the skilled oil worker earns on average 18 per cent more than general industry, the most experienced and knowledgeable professionals can enjoy salaries of 30 per cent more than their counterparts in other industries.

Hot bedding and T-bones for breakfast

The artificial nature of the offshore work environment was accentuated by the living environment – both shared the same location. No other workforce sleeps in the factory, or on the building site or down a mine – and mining is the occupation usually compared to offshore. Certainly no employee lives so intimately with such a highly volatile and sometimes unpredictable product. In conditions, accommodation, social life and catering, there was generally a marked contrast between rigs and platforms; the difference between frontiersmen and settlers. The semi-submersibles and drill ships were fairly rudimentary and basic, the fixed platforms more luxurious. It also depended on the fluctuations of the industry.

Sandy Clow regarded conditions on *Staflo* as acceptable. Nowadays they sound primitive. 'We had two- and four-man cabins. There weren't private showers but there was a communal wash area, not very well designed. There were no laundry facilities, so we used the steam heating lines; you cleaned out an old oil drum, filled it with water and then put the steam line in and washed the dirty laundry in that. It was a dirty job because the mud was diesel-based, nasty, and it stank. Your clothes really had to be washed out.' Michel Euillet worked on a rig called *Transworld 58*. 'The accommodation for service personnel was atrocious. Hospital beds, piled on top of each other, not even designed to be bunk beds. A single, bare light bulb, no curtain, and washing – the water probably came from the bilge, because when you took a shower you ended up with more oil on you. We had no cinema, but there was one of the very first televisions and videos. The videos, however, were mostly those that also can only be watched after hours. We had to buy the toilet paper. It was miserable. Hamilton Bros were giving the Aberdeen people a good name.' When painter and blaster Graeme Paterson first went offshore there were four-man cabins. 'Some places there were six-man, eight-man cabins and we were packed in like sardines. All they wanted was the guys out to do the work, never mind how they felt. But that's one good thing, they've changed the accommodation to only two men to a cabin.'

The practice of 'hot-bedding' became prevalent in the platform construction phase. Scots engineer Andy Lawrie, who had transferred from BP tankers to an installation manager on Forties in 1974, discovered this was the sleeping arrangement on the Alpha platform leading up to production 'There was a massive amount of men working on a platform designed to take eighty or ninety and we had one hundred and fifty. There were six men in a cabin designed for two. As installation manager responsible for health and safety and welfare, I considered the conditions less than satisfactory. They would go

on night shift and somebody would go into their beds. The recreation rooms were used as dormitories, maybe with fourteen or fifteen bunks in them.' It was not until 2000, under the implementation of new health and safety regulations, that the 'hot-bedding' practice was finally stopped.

Life on the *Thistle* platform in the Northern North Sea came as a culture shock for the Australian welder Ralph Stokes. 'The conditions weren't good. The rig's nickname was *"The Black Pig"* – and it wasn't nice to work on, it was pretty rough. The accommodation was pretty basic – modules set down on the rig outside – thirty two-man cabins, so bloody tiny it was unreal. We had sixty men to a block, and a toilet block with four shower cubicles, no privacy, no curtains and a gutter drain running through the showers. Five toilets, a urinal and six stainless-steel washhand basins – for sixty blokes; it wasn't very pleasant at all.'

There was a great contrast in conditions between the platforms and the earlier rigs. Radio operator Steve Russell-Pryce said accommodation on the platforms was good. 'But the rigs were rough and ready. In the newer ones, you had four or five toilets without partitions, so you sat talking to your mate while you were doing your business. After that they installed dormitories with two to a room. Quite a lot of the newer ones had showers. After work, you looked at the movies, although latterly we all had televisions in the rooms anyway. That was something of a "no no", socially; you came off, had something to eat, a shower, and then you lay on your bed and looked at a movie. That was life offshore.'

Conditions could be tight on the platforms too. Roy Wilson worked for years on *Beryl Alpha*. 'Everything was squashed into a little space. But I thought it was good. The rooms were for four people per room. As the company representative I generally got better accommodation. But it was mucky – standing in the rain, covered in oil and mud and working lots of hours. That was nasty, but we did it because we were so energised by the work. You might be out for 20 to 30 hours and only get a few hours sleep. Then you would get up and do it again.'

Old-timers still talk about the legendary quality and quantity of offshore food. Diver Keith Johnson said the Americans brought their own food from the States to Yarmouth and shipped it to the rigs. 'We lived like kings. Catering is nothing to what it was in the early days.' But he said that accommodation is now far better. 'They have single cabins, televisions; we had to sleep all over the place, get your head down wherever you could. Six in a cabin was nothing. Sleeping on the floor, whatever. Totally different outlook – totally different people. Today's guys wouldn't put up with what we had to put up with.'

Swede Lingard had the same opinion about the changed range of food. 'When you came off shift you could have anything from Dover sole to fillet steak – three, four times a day if you wanted, especially on the night shift. You could ring up the mess, "I just fancy a bit of Dover sole." Cook would defrost one and it would be ready at midnight.' Ronnie McDonald said on established platforms accommodation was good. 'Certainly comparable with your average merchant ship and the big thing was the food. Great efforts were made to ensure it was well above average and plenty of it. By and large that has gone – what do they call it now – portion management and cutbacks?' When Jake Molloy first arrived on Ninian Central, the food was superb – but not for long. 'That deteriorated quicker than anything else. You used to get steaks, one every meal – sirloins, fillets, T-bones. Lunchtimes, there was always a good seafood platter on, lobster if you wanted it. Then it was just steak on Sunday, then it started to disappear and finally, it was more steak mince than steak.'

Them and us

The practice of using specialist contractors on the rigs was a system brought to the North Sea by the American companies and became the norm across the industry. Hundreds of contractor firms would carry out the aerial and sea surveys and seismic, do the drilling, test the wells, perform the logging and analysis, provide the specialist engineering and construction, and, of course, the catering and the services. According to a number of present and former offshore oil workers a definite hierarchy has always separated the contractor or service employees and the 'company men'. That gulf was witnessed by diver Michel Euillet on the *Viking Piper,* a semi-submersible pipeline barge. 'I had been working above the deck and I had a good view of the difference between the cabin for the divers and that for the company personnel. Our cabin was large enough for four bunks, with one in the middle. There were metallic boxes for a change of clothes – eight in total for the crew change. Four of us could stand up in the middle, but there was no room for a table. The cabins for the American crew, each had an entrance, an office, a bedroom and a bathroom. The cabin for one man was five times the size of our room for four. I do not exaggerate.'

Neil Ferguson, a cementer for the major oil service company Schlumberger, believes there used to be a definite 'them and us' situation. 'When you were on the semis – the exploration rigs – the operators ran the show. They treated their own employees better than the service hands, who

were second-class citizens. If there was a shortage of chairs in the recreation room, the service guys had to get out. That was up to the 1980s and even in the 1990s. I was on a platform where a guy from the operating company came into the rec. room, staff keys dangling, went to a cupboard, opened the padlock, took out a newspaper and locked the cupboard. There were no other papers lying about, so I asked, "Do you mind if I read the paper after you?" "No, you can't." The guy read the paper then locked it up again. In this day and age – and that was a major operating company in the North Sea.'

The divide between the two occupying groups on the rigs probably united the contractors, according to Jake Molloy, who experienced some of the silly practices. 'The operators, the Chevron people, they had their own cinema and their own mess. The contractors had to go across to the old temporary living quarters and their own cinema. Somebody had to run back and forth to get the reels of film. Then they got rid of the old TLQs on the Ninian and we moved into well-kept two-man accommodation. The old demarcation was almost gone. Then I moved to Brent and I was right back into it again with Shell. It was the old, "You are not allowed in here. You can't have a cup of tea in here. You can't sit on these seats," and so on. It was a step back in time.' The power struggle in the cinema was a source of great amusement to medic Steve Russell-Pryce. 'The OIMs had their own seats. Well, the movie used to come on at seven o'clock and everybody was informed by Tannoy. So everybody who wanted to see it came down. Sometimes the company man, he would be in a filthy mood and he wouldn't come down until half-past seven, quarter to eight and say, "Hey, you've started the movie." So we had to stop the movie and rewind it. The company men literally were gods.'

But, as Graeme Paterson explained, in a competitively cut-throat business, service companies cannot afford to offend the operators. 'I worked on BP first, out on the Forties, and they were terrible. Their attitude towards contractors was absolutely horrendous. It was as if they never existed. They always thought they were better. Thankfully, that's all changed now. I noticed that, because after I left Forties, I then moved about the North Sea.'

The problem of contractor versus company was not always so black and white. Bruce OIM Andy Lawrie was frequently unimpressed by the managerial abilities of some contractors. 'You would have skilled men, engineers, welders, a lot of good tradesmen out in all weathers and they would see other staff working with warm outdoor clothes in warm offices. These guys had nothing like that. I told them to take it up with their own management. The reaction was "put on an extra jersey". So they came back

and I told them I would have a word. Within a month, everybody had proper weather gear and there were absolutely no further problems.'

But, for Michel Euillet, the most glaring example of the offshore 'class' system came one Christmas on Brent. The Alpha platform was attached by a 120-foot-long telescopic bridge to a semi-submersible service personnel accommodation block. 'It was an extremely interesting adventure to cross in bad weather, because one structure was fixed and the other mobile. The bridge was moving all the time and actually it collapsed later with two men on board. Anyway, the weather was bad and there wasn't a crew change. But we had a reasonable meal – for Christmas – served earlier than a different meal for Shell personnel. A television crew were filming the meal and their accommodation, which also was nothing like ours. The film was about how well Shell treated its staff. The microphone or the cameras were never put in front of us. The conditions were not that bad in comparison, but we were treated differently.'

The power change

That divisive hierarchy still continues, but in the mid-1970s there was a different shift in power offshore and it began when Graythorpe One, the first completed production platform, set out from Teeside, to take up the position it would still be holding thirty years on as *Forties Alpha*. Exploration continued to run in tandem, and in fact drilling activity was the highest it had ever been, with sixty-five wells bored, but by the end of the decade, most drilling was concentrating on the hundreds of appraisal and development wells. The production and development phase had begun and the driller was no longer king. As the platforms were put into place, drilling crews also began moving over to the new installations. This made offshore management 'more interesting' for OIMs such as Jim Souter. 'When BP had drill crews, collectively they were so unpredictable. Once we brought in contract drill crews that changed. You hardly knew they were there, they kept themselves very much to themselves. The BP crews, now, they were different. Many of them were from Trinidad, a couple were Americans, because they had the experience. And they were getting film stars' money. They were being paid in dollars at that time and it was more attractive.'

Swede Lingard, who had been working overseas, returned to Forties and spent a year as senior toolpusher on Bravo. 'That was different because there were a lot of people. I don't think it was such a good job, too many people sticking their oar in. Ninety per cent of my work had always been on wildcat

Plate 27.
The tragic end of a great North Sea lady – the *Sea Quest* reduced to a shell by fire while operating on the West Coast of Africa.

jobs. The toolpusher, if he said he wanted to do something, everything else had to wait because they wanted the oil out of the ground. When I came back as relief drilling superintendent in the 1980s, I just couldn't believe it. They were producing the oil and production came before drilling. They were getting the money out then.' Swede is now retired and has returned to farm life in Lincolnshire, where he rears game birds. His great friend Little Joe had tired of drilling and trained as a mud engineer. 'I was on all of the BP platforms at one time or another. One thing didn't please me was when you were working on a platform there was actually drilling going on and the wells being drilled were also being produced. The production teams took priority. We weren't used to that and it took a lot of swallowing. We were the boys who had discovered the oil and these guys had just come out on the production side. And they got the best cabins.' Joe is also now retired.

'Drilling was always the poor cousin of the platform side,' claimed Martin Reekie, who had joined the industry by the time the new regime was fully in control. 'We were the people who made the place dirty. They were bigger outfits, of course, especially the Brent – huge platforms and many hundreds of people more than you had on the semi. It was quite different.' Mike Waller, who was also a driller, was aware of the shift in the offshore hierarchy. 'They were production people, a completely different branch from ourselves. On the drilling side, more rigs were beginning to work for us. We were starting to have a lot more activity. There was a buzz about these times you know. On the shore side, there was a whole new world arising. You could see the importance of it for Britain and there was a lot of pride in it as well.'

The Shell men were describing the end of what could be termed, the first Oil Age. In 1975 another signal was the departure of one of the original trailblazers, *Staflo*, which was replaced by *Stadrill*. The rig, rechristened *Petrobras XXI*, has been converted to a production facility in Brazil. Her great rival, *Sea Quest*, finally left in 1980, but sadly was destroyed in a fire off West Africa. Almost coincidentally as *Staflo* disappeared over the Atlantic horizon, the rough, tough men from Louisiana, Oklahoma and West Texas, who had set the great enterprise in motion, were also beginning to depart and a new, more productive era had begun for the Scots and their fellow oilmen from across the marine border.

[4] The Good, the Bad – and the 'Coonasses'

The woman in the small Nottinghamshire village of Eakring, on the fringes of Sherwood Forest, was highly suspicious of the men with unfamiliar accents riding huge trucks laden with heavy equipment into the depths of the pretty little copse known as Duke's Wood. This was 1943. The country had been at war for four years and everyone knew to be on the alert for the appearance of strangers in the community. Finally, she stopped one of the men. He was tall, tanned and polite. She asked him, 'Who are you and what are you doing here?' The man drawled, 'Why, ma'am, we are Americans. We're from Hollywood and we are making a movie hereabouts. We are just waiting for John Wayne and the rest of the cast to arrive!' The woman was happy to accept that bizarre explanation – for a time.

Later she and the rest of the people in the area learned the real – and just as implausible – reason these Americans had entered their lives, but they never talked about one of the best-kept secrets of World War Two. The strangers had been invited from the oilfields of Oklahoma to Britain to find a reliable source of the fuel desperately needed at a crucial stage for the vehicles, vessels and aircraft of the country's hard-pressed Armed Forces.

The remarkable story of the American civilians who came to the aid of Britain, some two years before Pearl Harbour forced their country into the conflict, was only made public a few years ago. Their historical importance now is that they were the first to teach British oil workers how to drill deep wells – the vanguard, in fact, of the hundreds of their countrymen who were to launch, twenty years later, the United Kingdom offshore oil and gas industry. A secure supply of oil based on the homeland had become crucial and it was known that there were reserves of fossil fuel in the Nottingham area – the first commercial UK well had been brought in at Eakring in 1939. However, there was neither the equipment nor the knowledge to operate it and exploit these reserves; that existed in America. At the Government's behest, Philip Southwell, boss of the Anglo-Iranian Oil Company – and later chairman of British Petroleum – flew to Oklahoma to talk to the owner of the foremost drilling company, Noble Drilling. In a totally altruistic gesture, the

Plate 28.
The forty American drillers who came from Oklahoma to Sherwood Forest in 1943 to produce oil. Drill superintendent Gene Rossiter is the man without the hard hat. *(Nottingham Evening Post)*

company agreed to send, at their own expense, forty of their men and the necessary equipment to Britain. The crews crossed the Atlantic on the *Queen Mary*, their drilling gear transported separately on naval vessels. One ship was sunk by U-boats, so they sent another. The oilmen were set a target: produce 100,000 barrels of oil in 365 days, a figure which was well exceeded. Unfortunately, the operation was marred by death when an oilworker plunged from a derrick, the first recorded fatal accident in the UK oil industry.

Dennis Shepherd worked in the laboratory at Eakring straight from school at the age of fifteen and he was there when the Americans arrived. 'I had to pick up well and tank samples to be tested for quality and quantity. I got on all right with them and I used to go to their canteen. Taken all round, they were OK. They were actually billeted in a monastery, but they got on all right. Can you imagine it – all these tough American oilmen living in a religious establishment? It is an amazing thing, we were never sworn to secrecy, but nobody ever talked. It was vital that the oil was kept safe from air raids. The project was about absolute survival. This country was on its knees. Then in

THE OILMEN

Plate 29.
The impressive memorial in Duke's Wood, Nottinghamshire, erected as a tribute to the American drillers of 1943. The style of the statue of an oilman is remarkably similar to the Piper Alpha Memorial in Aberdeen. *(Nottingham Evening Post)*

1943 this little oilfield produced 110,000 tonnes of oil – it was our salvation.' The 'little oilfield' lasted another twenty-three years, providing a total of 4,500 barrels a day. The job done, the Americans went home. Recently, some of them returned for a ceremony to mark the unveiling of a bronze statue, The Oil Patch Warrior, set in the lovely surroundings of Duke's Wood, where a number of the oilfield's now redundant 'nodding donkeys' – the familiar land production pumps – have been retained. A replica was also erected at

Ardmore, Oklahoma, home of Noble Drilling. There is a haunting similarity between that embodiment of an American roughneck and the figures of oilmen in the Piper Alpha disaster memorial in Aberdeen.

THE OILMEN

The impact of the successors to those American wildcatters, who arrived in the 1960s to look for hydrocarbons in the southern North Sea just as the Nottinghamshire reserves were finally drained, was rather different. They are remembered with far more mixed feelings than that of the young analyst from Eakring. Peter Carson of Shell claims, 'If you talk to many of the earlier guys who worked on rigs as roustabouts and roughnecks on American crews, you will find a huge animosity.' But like their wartime colleagues, those American drillers and toolpushers are now recognised as having been vital to the progress of the industry in the southern seas and later in the northern sector. In a paper for the Petroleum Exploration Society of Great Britain (PESGB), Colin Fothergill points out, 'It is sometimes forgotten that quite a few pioneers of exploration in the early days of the North Sea were independents from North America, who had only limited experience overseas, but who had years of offshore experience in their own back yards . . . The know-how brought to the UK by American oilmen enabled exploration and drilling to get off to a flying start in 1964, after the first licences were awarded and ensured the high level of activity which followed.'

In another PESGB paper, Nigel Anstey also underlined the transatlantic drillers' importance. 'In the early days, most of us had learned our trade from Okie geologists, Texan doodlebuggers and Louisiana rednecks. Make no mistake, those guys were good. They were hoary with experience and had a good practical way with them. Oil exploration was an American skill; the jargon was American, the literature was American, the equipment was American. But after years of saluting the experts, there comes an itch. To hear someone say, "Well, that sure ain't the way we do it in Texas, but I guess its OK."' Not every North Sea veteran shares that view. Nor was the reaction to the British crews always so amiable and equable. Especially not from the early drillers and toolpushers who called themselves 'coonasses'.

For about a decade, through the frantic 1970s, apart from BP and Shell, these Americans led the hunt for oil. By the 1980s, as production overtook exploration in priority, most of them had gone back to the States, replaced by compatriots who were a different breed, academically qualified and cultured family men who operated on a different level in their companies. Offshore the Brits had taken over from the 'rednecks' and the 'coonasses'. But those highly colourful first Americans left behind a legacy of anecdotes which told what they were like, how they treated their crews, and how they lived, worked and played the only way they knew how and that was hard and rough.

THE OILMEN

Alan Higgins, ex-OIM on the Ninian platforms and a former general operations manager in Aberdeen for Chevron UK, drew a graphic word picture of them. 'My direct boss was a great big cigar-toting American. Billy John Kilpatrick was his name. He came from Louisiana. And there was another one called Red Ruffian. They used to go down Union Street in Aberdeen in buckskin jackets, cowboy boots and Stetson hats. If you walked 10 yards behind them, it was a joy to see the faces of people coming towards them. They always had cigars in their mouths but they never smoked the cigars, they just chewed them.' Similar Americans drilling in the southern gas fields frequented the hotel in Cleethorpes where Little Joe Dobbs served the wine and he became friendly with them. Offshore, he found they were a different breed. 'I thought they were all characters, but when you were working they were hard guys. It didn't matter whether you were friends or not. A couple of times I was physically threatened. It was dangerous on the platform in those days. We had to go round and set up this mooring equipment without safety belts, and I refused. They said, "You'd better get off on the next helicopter." That is the way it was – hire and fire because they could get as many men as they wanted.'

American toolpusher Gordon McCulloch was a dominant figure in the career of Charlie Anderton, a directional driller with Santa Fe. 'I was Gordon's blue-eyed boy. So I said to him, "I want up the derrick." And he said, "I will have to take you to see your rig superintendent." So I went to see this lad, a real coonass, and Gordon said to him, "Charlie said he is ready for up the derrick." The boy said, "How long have you been on the rigs?" I said six months. "Well, son," he says, "there are two ways of doing this. You can either jump on a ladder halfway up and fall off, or you can climb the ladder slowly from the bottom until you reach the top." And he turned back to whatever he had been doing. So I said to Gordon, "I take it that was a no." Mind you, I have always been very pro-American. I liked them. I liked their mentality, the go-getter mentality. If you can cut the cake, you get to eat the cake. That suited me. I remember I had flooded the pump room twice – it was a foot deep in mud, and the toolpusher came down and he looked at me, "That will be the last time you do that." I said, "Yes, that will be the last time." Next time I would be out of a job. I liked that and I still believe men work better with a little of the fear factor. In the present regime in the UK we don't have that.'

Equally impressed was another farm boy, Swede Lingard, although the first days offshore were a culture shock. 'I remember this toolpusher, he said to me when I started, "Where Ah come from, the further down the street you go the tougher they come. And Ah live in the end house."'

John Carter watched the Americans in action from the wheelhouses of the vessels he skippered in the North Sea, including tug supply boats servicing rigs from the States. 'It was an eye-opener for me working with those people – people with no marine experience at all, from Texas or Louisiana. They were totally ruthless with our people, and very gung ho. "We'll cool this m— f—ing pond." Or, "We'll show these spaghetti-munchin' bastards how to lay pipe." That was kind of the attitude. The Norwegians had the same problem, but they just didn't put up with it. They told them, "You are not treating our people like this." Whereas our people turned a blind eye to it.'

Remote underwater vehicle technician Derek Stewart said, 'The drillers were mostly American in those days. You would go to the mess for something to eat and they would say, "Hey, what you doin' eating? Ain't you fixed the damn TV yet?" Tough people and it was the first time they had been working in this area and they didn't treat us very well. You were "white niggers". That is how they started off. Most of them had been working in West Africa or Nigeria and the local labour took that kind of stuff, like. We didn't here. We sorted them out. Most of them left and there are hardly any here now.' The pressure the early American toolpushers exerted on divers to take risks worried former naval diver Commander Jack Warner, later Chief Inspector of Diving. He wrote in his book, *Requiem for a Diver*,* 'I know that in many instances, and suspect in many others, diving supervisors were co-erced into ordering a dive to start against their better judgement. Back in those bad old days, the toolpusher was God on the rig; he was invariably a tall Texan with high-heeled boots ("shitkickers") and Stetson and he had a favourite instruction to divers, "Get diving or get your ass off this rig." And he meant it. To him, Brits, and indeed other foreign nationals, were nothing more than "white wogs". They were there to work or catch the next chopper to the beach, because he said so. They were labour with which to complete a drilling programme, nothing more. That attitude has, fortunately, changed considerably.'

Former Chevron OIM John Nielsen found a different reaction in his experience of American-type management offshore.'If they found they could get you on the run, they would get you on the run, because that was their style. It still is. You turn round and bite or snarl, they will back off. The guy in overall charge during the commissioning of Ninian North, Gene Harrison, he was called – he was quite famous. I got on with him all right. We had a problem and I had just taken over from my "back-to-back" when I got a

---

* Warner, J., and Park, F., *Requiem for a Diver* (Glasgow, 1990).

bloody great bollocking through a message from Gene. I sent it back and said, "It has nothing to do with me. I am out here to build a platform as quickly as possible, as cheaply as possible and with the least amount of problems." He sent one back, "You are quite right, John. You have to build it with the least amount of money and despite the problems." So he had a good sense of humour in trying to defuse the situation.'

Charlie Brown, among other things a former OIM with Shell, recalled a Louisiana toolpusher he worked under. 'This guy, he could make hole, but don't ask him to write his name. That is how petroleum engineers came to be the kingpins of the industry, because rigs used to have a pet engineer and if you go back far enough, you will find he was the only guy who could write. But those toolpushers were good. A lot of them came from the land rigs in Texas. Moving from one area to another they picked boys leaving school to teach and so it went on. Until they got into the offshore business, Americans would not go what they called "foreign"– they stayed at home. You then got the kind of guys who couldn't get the kind of jobs they wanted back home.' Before Jake Molloy became secretary of the oil workers' union, the OILC, he worked on a range of offshore installations. 'There was this old guy, Elmo Reid, I have a copy of a risk assessment document and he had just signed it with an "X". All he had ever done in his life was wildcat. To me, he was a good old guy, but to his crews he was a beast. He would wander about with the old denims on and a flat denim hat. He never caused me any grief, but when he spoke to his crews, they hated him with a vengeance.'

In the early days, Americans more accustomed to the Californian coast and the Persian Gulf found the weather hard to handle and for the dry land drillers – like the British – it was an education. Little Joe remembers their first reaction to the weather. 'They pooh-poohed it initially – that and the sea conditions. But then they became aware of it. The majority went back thinking that this was the worst place they had ever drilled in. They were actually very good and they knew what they were doing on the drilling side. They were teaching everybody else and we were all gaining experience as we went along.'

Many of the old-style drillers from the States had difficulty handling the UK's offshore legislation. Ex-Chevron OIM Alex Riddell said, 'And this was before *Piper*. They couldn't understand these Department of Energy inspectors coming on board, telling them, "You must do this and you must do that." They had the equivalent kind of inspectors in the US, but they didn't have the same powers. Neither could they understand the Lloyds inspector, when he said, "Right, I want that crane shut down and I am revoking the certificate." They couldn't handle that. So a lot of them said that enough was enough. The American management ashore, now, they could handle it, but sometimes they were very reluctant to accept it.'

Safety was the issue that almost brought two stubborn Americans to fisticuffs late on a stormy night in the North Sea. One was a barge master, the other the construction boss on a platform, linked to the barge by a bridge. Ronnie McDonald was there. 'This bridge had a winch and pulley arrangement which never worked properly and had been completely non-functional for three days – a fact kept quiet. The weather was deteriorating rapidly and there were 800 men on deck, trying to get off the platform. Others were standing on the barge refusing to cross the bridge. By four in the morning, it was clear a catastrophe was in the offing. The barge master and the construction superintendent were screaming at each other above the noise of this Force 9/10 gale. The super wanted to keep the bridge in place to get his men over and working. The barge master wanted the bridge lifted to lie off the platform before the anchors pulled his barge over. I don't know what happened first; whether the bridge actually collapsed into the sea or the barge moved away, but it was all more or less simultaneous. There was wreckage falling, block and tackle flying everywhere and people running for their lives. There is no doubt that men would have been on that bridge if they hadn't refused to cross it.'

Alan Higgins found that Americans loved giving good news and credit for good performance. 'But they had a greatly different view on bad news and poor performance. In America there's no such word as failure, you know. It might be stress, but it's not failure, and we had the same problem.' Diving manager Keith Johnson reckoned it was really all bluff. 'I found out when I went to work in the States it was kidology, really. I was in Louisiana and I thought we had to provide a service, seven days a week, 24 hours a day. Then I realised that over there they all went home Friday afternoon. So I did what we do here – in on Saturdays and Sundays, and within two years I had 80 per cent of the business.'

Straight from university, a very different work culture awaited young engineering graduate, Roy Wilson. He was stunned by the rough sense of humour of his superiors. 'They were the real "good ol' boys", tobacco-chewing and spitting rednecks, and the language to go with it. There were two toolpushers on the rig and they astonished me with the things they were doing. One was very good and the other was not. One was very much in command and knew what he was about, the other one did not. They used to play all sorts of tricks and made a fool of you all the time. I can recall getting a bit depressed because they didn't appear to have any respect for me – a young guy straight out of university. Eventually I mentioned it to one of the

THE OILMEN

assistant toolpushers, and he said, "Roy, they're teasing you, 'cos they like ya. The minute they stop teasing you and making fun of you, then you know they don't care for you. That's just their way of showing you they like you."'

Graduate structural engineer Jane Stirling's experiences of working with Americans offshore with Occidental present a different slant from some of those stories. 'The drillers were far punchier – yes, definitely. And I was far more wary dealing with the drillers. I worked with a really great drilling manager at the time, an American, and he would never treat me differently. He was a good guy. He would always say to me, "If they ever give you any trouble, Jane, you just tell them they have me to deal with." I always thought I got good back-up. Technically, however, if I went to a drilling manager, although he'd got big boots, a big person, lots of power, he would always listen because he knew fundamentally I could tell him information that would allow him to do something or on the other hand prevent him from doing something. But I suspect a lot of men and women would be put off because they were pretty hard on their crews. I was a lot younger, and maybe it was Southern courtesy, but they always called me "ma'am" and were polite, yet I know they were very tough.

That is not what medic and radio operator Steve Russell-Pryce personally encountered with some of the original coonasses from the Deep South. The reason is that, in spite of his very English double-barrelled name, he is West Indian and black. 'Going offshore was daunting at first. When I started, they couldn't get enough people. You saw an advert or something and you sent a CV and they would phone you up and offer you a job. Because of my name, they didn't know what colour I was, so they would tell me to go to the heliport. Once, this agency chap, who brought my safety gear, walked up and down the line, saying, "I'm looking for Steve Russell-Pryce, a medic." He never even looked at me because he thought, "He's not one of mine." But I said, "That's me," and he said, "Oh, my God." I said, "Is there a problem?" He said, "Oh no, oh no." So I went out to the platform, and the company man, an American, was standing outside the radio room when I came off the helicopter. He sort of pointed to me in the mass of people and pointed back to the helicopter. The guys knew exactly what he was doing and it was like the parting of the Red Sea. I walked forward and I pretended I didn't know what was happening. He said, "You, boy. Get back on that helicopter. I don't want any nigger on my f—ing rig. So get back off." And he said to the radio operator, "Don't let that helicopter go. Get him back on." That is what I had to do. And it happened to me about four times in different places. When they changed the law on racial prejudice, I started to use that, and I kept saying to the Americans, "You see, this little island, it is mine. You are not in the States any more." I had to become a smartass, you know. I had to sort of give as much as I got. I believe I was the only black man working in the North Sea at that time. They used to say that we were bad luck on the drill floor. Well, that was their excuse, anyway.'

American reaction to these stories varies. One company boss said that he had never heard of such situations, adding that there were probably instances where 'you had some hard-nosed Texan who got pushy but somebody below me would do something about it before it ever got to me. I had people who weren't going to tolerate that.'

Working for Conoco, an American company and beginning on the drilling rigs, Roger Ramshaw learned to understand what they were all about. 'The whole atmosphere was wild, but it was fun. You had these characters you saw in John Wayne movies. That was fiction, but they were real. It was very much a "can-do" society. Physically, you had to be a tough guy. Safety was talked a lot about, but it was a different order of magnitude to what it is now. But it was by and large a meritocracy. I found the Americans were more comfortable than the British in terms of "what do you do – not how you speak".' Other UK oilmen were amazed by the 'can-do' culture of the Transatlantic companies. John Wils, who eventually became director general of UKOOA, left Shell to join ARCO, one of the second-generation American companies in the 1980s, to work as a production manager in the southern North Sea. 'The contrast in culture of suddenly working for an American company was really quite interesting – very different. I remember asking my boss in London, "Don't you think you need an office down in Great Yarmouth?" He said, "Yes, you better go and do something about that." I said, "What do you want me to do?" And he replied, "Decide what you need, go talk to some developers, choose the proposal you want and go ahead with it." So I found myself travelling down to Great Yarmouth with a team of surveyors to set it all up. With the Americans, you had to deliver, and if you didn't, that was it.'

Some Americans themselves accept the fact that the movie tough-guy style would have been difficult for their crews to appreciate. Tom Marr, a project manager on Conoco's Murchison and Hutton operations, said he had no experience of it himself, but it wouldn't have surprised him. 'Drillers are a funny breed of cat anyway, renowned round the world as a wild bunch of idiots. They usually call a spade what you would expect them to call it. They think they are the only ones that can walk on water. I am not surprised the workers would feel as they did. These guys were steeped in many years of slinging the bullshit around.'

His colleague Jack Marshall believes that the way early toolpushers and

THE OILMEN

drillers managed their crews was 'absolutely necessary'. He comes from an oil family, and from the age of fourteen worked in the oilfields of Oklahoma as a roustabout and a roughneck. Eventually, he graduated with a Bachelor's degree in Petroleum Engineering from Tulsa University and joined Conoco. He was part of the company's determined drive from a small Mid-western oil company to a large international business, as vice and then executive vice president of international exploration and production. 'Eugene Rosser, the leader of the American crews who drilled in Sherwood Forest during the war, told me one time that when he turned in his first report and had made 1,200 feet, the oil company called him a liar and said, "Nobody has ever made 1,200 feet the first day." I am sure the first UK roughnecks working on drilling rigs would have been pretty shocked. But without meaning to be really ugly, one thing they needed to understand was that when you signed on, you worked, period. That was just what came with it – effort. It was just a difference of philosophy and it was absolutely necessary. You got a drilling rig costing you $300,000 a day. If you run it right, you get x production, and if you don't, you don't get near x production. I used to tell folks in the UK and in Norway, "You are letting us in on your oil wealth, which you own – we know it – but you are letting us in on it because of something we bring to the table." One of the things we brought to the table was the ability to drill wells better than anybody else in the world. For the UK offshore workers it was a new world. They didn't understand it. They had never been there and they had certainly never been there under our management style. But, gosh, I found out in the construction stage, the development stage, that they had learned to appreciate it, joined right in and became excellent.'

He remembers the North Sea as 'one of the most exciting times anybody in the world will ever have. I think I lived in a time when we took the centre cut out of the watermelon. We did more things, we had more fun, we had more challenges and I know we enjoyed it more than anybody today is enjoying it. We approached it with a lust for life and the desire to show people what we could do and why we did it that way.'

It would be fair to say the Americans weren't the only unforgiving foreign taskmasters. Rigs operated by Shell, such as *Staflo*, were headed by Dutch toolpushers and drillers. Peter Carson said they were also very tough and very abrupt. His first time on a drilling rig was in the Hague on an induction course in 1971. 'We had a "kick" and we were weighing up the mud and there was all sorts of shit going on and I had no idea what it was all about. We were loading barites – things were blowing off and hopper covers were blowing off and this old toolpusher was swearing away at me in Dutch and kicking me up the arse, "Get over there and do that." And we did

Plate 30.
Jack Marshall, former Conoco executive vice president of international production and exploration.
(Conoco Phillips)

it. That was the start, they made you learn from that. I don't think they were ever vindictive. They were very dictatorial, whereas the Americans were very brusque and macho.' His former colleague, Sandy Clow, who began work under drillers who had learned their trade in the gas fields of Groningen, is far less diplomatic. 'Some of the old Dutch drillers and toolpushers weren't very nice at all. A bit excitable. I remember in Middlesborough there were two Dutch toolpushers in charge on different weeks. They didn't like each other. Now, everything was being painted in the colours of the various companies involved in the rig, but one of these toolpushers had a different idea. He decided everything had to be painted lime green. So, we got out the paint and the rollers and away we went. The next week, the other fellow comes on. He was shocked, so out come the paint brushes and everything had to be painted back to red and blue. That actually happened a couple of times until big Jock Munro questioned the amount of paint being used and that was the end of that.'

Gradually, as the industry developed with more rigs and platforms, and as the workforce expanded rapidly, acquiring oilfield skills themselves, the old-style Americans began to return to the States. Alex Riddell said they had begun to find themselves out of their depth. 'Once the accountants started asking for budgets to be put together and you handed accountability to the guys offshore, they just couldn't handle it. So they said, "Right, give me my airline ticket back to the States."' The Americans who remained were another breed altogether. 'They were guys who were going to move up in the company, people-type people who could communicate, a lot younger, more vibrant – and they could adapt to the UK culture. It was a good change and did a lot for us.'

As British oil workers began to slip into positions of power on the rigs and platforms, unfortunately some picked up bad management habits along with their new skills. One Sunday morning on Ninian North, the crew had been called to muster for lifeboat drill by the services supervisor. But the drill wasn't running smoothly; in fact it was chaotic, and the supervisor was apoplectic at the crewmen's unsuccessful attempts to board the boat. The problem was fitting fifty people into a small boat. There was also another very big handicap: the presence of a veritable giant, probably the biggest oil worker ever to go offshore, 6 feet 5 and reputed to weigh in at 32 stone. He needed two or even three seats. But the crew were trying, installing the bottom tier of men first and then building around them. The supervisor was still unhappy. He thought the crewman counting in the people was too slow. So he started to shout at him. From out of the crowd came the call, 'Piss off'. The supervisor lost his temper and demanded to know who had spoken, but

THE OILMEN

no one would own up. 'Right,' he said, 'everyone up on the walkway.' He took their names and informed the company they were all sacked and would be run off on the next helicopter. Later he climbed down. Offshore plumber Jake Molloy, who was the slow number counter, said, 'It wouldn't have been a first; people were being run off almost on a daily basis. That guy was just a nut case.' More significantly, the supervisor was also what could well be called a 'replica redneck,' a local man who had adopted his American bosses' unique if doubtful style of man management – and also their accents. Jake said, 'It was, "Take the freeway to Stonehaven," and it was sad because they always let it slip. "Goddamn, you sonsabitches. Why don't you haul your ass up there an' ging an' get a dishtool an' clean up 'at mess." Probably shooting that line was what they believed would keep them their jobs.'

Medic Steve Russell-Pryce was another who maintained that the Brits who had risen through the ranks were just as bad. 'Oh yes. They chewed tobacco and they tried to be like the Americans. Can you imagine somebody from Aberdeen or Turriff cursing like a redneck? I think they were like this in order to survive. Most of those bosses were all right actually, but some of them could be just pure bastards. They were just unbelievable. They would scream and scream – which I never found was the best way of getting work out of men.'

In the final analysis, the early Americans have to be understood against the background of the enormous pressures the industry was under to bring in the oil and gas. The oilmen from Oklahoma, Texas and Louisiana were as they were because of what they were, where they came from and what they had to do. They knew no other way. There are uncomfortable parallels to be drawn from the participation of Scots in the colonialism of the British Empire. In a perceptive aside, Chris Harvie suggests in his history of North Sea oil, *Fool's Gold*,* that 'we were no longer, however guiltily, the exploiters, but the exploited'. And he asks if we regard 'the barbarians – whether Texas oilmen or City dealers' any differently from the way 'the Brahmins had regarded the Scots soldiers and civilians who took over India'. It is an indisputable fact, as Nigel Anstey realised, that the Americans were the key element, the catalyst, of the complex economic and industrial equation that inspired the 'colonisation' of the North Sea oil province. In drive, know-how and get-up-and-go, they were the oil patch warriors of Duke's Wood reborn.

---

* Harvie, C., *Fool's Gold. The Story of North Sea Oil. How a £200 billion windfall divided a kingdom* (London, 1994).

A Breed Apart [5]

The Italian pipe-laying barge was anchored up off the north-east coast of Scotland in foul weather, sometime in 1974. The seas were very bad, too rough to lay. The stinger, the curved gantry down which the pipe was delivered from the end of the barge, had been lost overboard. The supervisors wanted divers to attach a line to the end of the pipe, but they were refusing to go. The water was too choppy and it wasn't safe. But they just kept offering the divers more and more money until they reached £500, a lot of money in those days. One of the divers said, 'OK, I'll take it.' Brian Porter from Derbyshire was the BP inspector, the company man on board the barge. 'I can see him now. They took him out in a little boat like a rowing boat. One minute, you are looking down at it. The next minute, it is above you, the sea was that bad. The diver goes over the side and just disappears. But he got the line on and he lifted it back up. He got £500 for that. Some of the other Italian divers were watching. One of them said to me, "I work to live, I no live to work." He couldn't understand how anybody would do it for that.' A small vignette which typifies the early divers of the North Sea oil and gas industry who were willing to tackle the most perilous jobs offshore for the greatest rewards. Their true value to the industry was summed up by a former Naval diver, Commander Jack Warner: 'It may seem an old hackneyed phrase that without the divers there would be no North Sea oil, but in the 1960s and 1970s, it was absolutely true.'

Like the drill crews and rigs, the first divers had followed the hunt north from the southern North Sea. By the middle of the 1970s, there were more than a thousand operating in the northern sectors. They were undoubtedly a race apart, viewed by those who employed them and worked with them with a curious mix of awe, admiration and amusement. Chevron's Alan Higgins thought they were 'a lot of brave people'. But, he added, 'they were a real harum-scarum lot. I mean they were skilled, but they only got paid big money when they were actually in the water diving. They spent a lot of time in saturation tanks and living in terrible conditions. They were probably, along with the drillers, the cowboy kind of element. Some divers were always

Plate 31.
Brian Porter began working on land gas pipelines in the south of England but later served for a spell as a BP representative on the indomitable pipeline barges of the North Sea.

THE OILMEN

looking for the shortcuts and believed the rules didn't apply to them.'

Charlie Brown from Shell compared divers to welders, 'They would find a thousand reasons not to do a job – but they would never give any reason why they should do it. Wild – this was the myth and the aura they spun around themselves.' On *Stadrill*, the driller Mike Waller regarded them as something special. 'A different breed, like an elite, especially when they were saturation diving. Not a nice thing, sitting in the decompression chamber. I remember looking through that little glass window at the poor devil in there. When I started to get involved in planning their dives, I gained a greater appreciation of what those guys were doing. I have heard about a cavalier attitude – but I didn't find that with the ones that we had – you had a lot of serious guys.'

There were always divers based on a barge during the Forties hook up, when Jim Souter was an OIM. Although they were not his responsibility, he was concerned about their habits. 'They were an amazing bunch in the early days, but for guys so obsessed with physical fitness, a lot of them resorted to chemical abuse to keep going – "uppers and downers". I found their habits disturbing. They were also supposed to work just twenty-eight days and take leave, but some of them came off the helicopter at Dyce and immediately signed on again as divers' helpers and did another twenty-eight days. Crazy.'

One of their own, Jim Limbrick, naturally had a totally different view in his book, *North Sea Divers: A Requiem*,* 'These men were the salt of the earth and prepared to work long and hard in conditions that they had no reason to believe would cost them their lives. In those days, it was all push to get the oil out and the "toolpushers" and the diver pushers were God and had to be obeyed.' Former Naval flier Dick Winchester piloted manned submersibles, which were employed extensively for diving operations in the 1970s. 'I found it personally extremely satisfying, it was just wonderful. We always used to joke that the company gave us all these toys to play with and all this wonderful food and time offshore and they paid us as well.' One diver has described what he did as 'just like taking a bicycle to the factory'. Diving company boss Keith Johnston explained 'What a lot of people don't realise is that diving is just a way to get to work – it's what you do when you get there that counts.'

But first the diver had to get there and his route to work was through unlit murky depths of deathly cold water, prey to sudden powerful surges and treacherous currents. In the early days, dives could be as deep as 400 to 600 feet and a diver only has to reach 33 feet to be subject to twice the

---

* Limbrick, J., *North Sea Divers: A Requiem* (Hertford, 2002).

Plate 32.
A breed apart – a North Sea diver at work in the murky waters deep below an oil installation.
(Subsea 7)

normal atmospheric pressure on the human body. This increases every further 33 feet, and it is the pressure that causes the danger. In the 1970s, the bounce system, involving bells and decompression chambers, was used. The deeper and longer the dive, the more time was needed for the body to adjust to the gradual change in pressure – decompression. Saturation diving, which involves descents of 1,000 feet or so, means the diver is subjected to such pressure his body becomes saturated with the mixture of oxygen and helium gases supplied from the surface. The significant factor is that whether the dive is 12 or 24 hours, the decompression time remains the same. The system was to enable dives of greater depths and longer time scales, spending weeks, sometimes months, in saturation under pressure. There was also, in the early days, the ever-present threat of hypothermia. Despite the purported 'aura and myth', there was very little glamour. 'I have spent days and nights frozen half to death,' says Jim Limbrick, 'covered in oil, inside as well as outside of a wetsuit, for the best part of a week on a grounded tanker; dragging out dead bodies; underneath grounded ships in the pitch-black dark with screeching, grinding steel all around you; breaking ice on winter days to get into the water and tossing about on storm-driven dark oceans with half of it inside the boat.'

Most of the early recruits came from Naval backgrounds. One such was

THE OILMEN

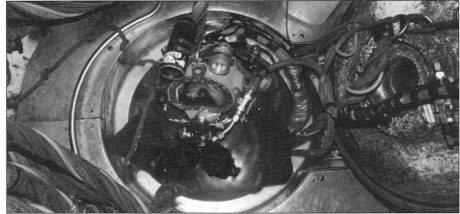

Plate 33.
A diver prepares to descend; in the bell and welding a pipeline. *(Kenny Thomson/ John Greensmyth, Technip Offshore)*

Keith Johnson, who was the first diving consultant to arrive in Aberdeen. 'I was a navy diver and I came out as the gas fields were opening up off Yarmouth. My class was the last to do standard diving, and in the beginning it was diving suits and diving bells. There were good jobs for divers in those days – £100 a week in 1967. That was unheard of. I was earning £7 a fortnight in the Navy. I was working for a small company, Marex, and I came up to the North Sea in 1968 on the *Sea Quest* for BP. It hadn't found at that time. Then it was the *Staflo* for Shell, *Gulf Tide*, then *Glomar V*. I commuted from Yarmouth and I was the manager when Comex bought the company out. Eventually, I moved up to Aberdeen.' In the southern sector divers used heavy suits and there was surface demand swimming gear, scuba, because of the shallow water – about 160 feet. 'Up here it was deeper water and it was all mixed gas, helium and oxygen and diving bells. Divers were used to make sure the BOPs on the seabed were in position, and we were also involved on inspection, trenching and monitoring pipelines. In those days, they also welded pipes spooled off the barges. The divers were the eyes underwater. I can't remember ever being worried. Everybody just wanted to be there, they were so keen to get in to it because it was more exciting than now. It was a developing and growing business – now it's all engineering and accountants. The last thing anyone bothered about in those days were costs. It was just – "Get the job done".'

Frenchman Michel Euillet first came to the North Sea as a diver in 1972, via Lerwick in Shetland, when still only in his twenties. Michel, now settled in the North-east, is an intelligent and perceptive observer of the North Sea operations. He started in the oil industry in West Africa on supply boats; he was also an engineer and a qualified commercial pilot. 'I was producing picture postcards and I used rather questionable small airstrips. The authorities thought I was a spy. I was twenty-six at that time. So I went back to France and passed my maritime qualification, which also meant I could teach diving. That was in Marseilles in the winter of 1970–71.

'The way to make a pile of money in diving was to become a petroleum pipe welder, so I took a course in welding. Eventually I joined a company in Paris which specialised in submarines. I only dived once – but I maintained the video and communication gadgets. Eventually, we got a small contract with a Canadian firm, KD Marine, who were moving in to the North Sea as a diving company. I installed some of the earliest guideline underwater television, the first on a semi-submersible, *SEDCO F*. The rig was working for Shell – off Shetland. There were no fields at that point. I knew quite a lot about the diving industry, about submarines, and I had also seen jack-ups in Africa, but I knew next to nothing about deep water. I had never seen

THE OILMEN

Plate 34.
In the eery half light of the subsea world a diver leaves the safety of the bell to begin a saturation operation. *(Subsea 7)*

floating platforms, so it was a bit of a shock, but I was the same as everybody else – it was new to everybody.'

Keith Johnston had left Comex and set up on his own in Aberdeen. 'I was the only diving consultant at that time. I looked after all the oil companies, giving them my advice. They were struggling; they knew very little at that time and we were on a fast-track development. Saturation diving was only just beginning and we were diving out of bells by then.'

Both men described the 'United Nations' in the diving industry at that time. Michel said, 'There was a certain number of Italians, they had appeared fairly early. There were quite a lot of Canadians, Americans (mostly from the Gulf of Mexico) and New Zealanders, but very few Australians. Funnily enough, there were quite a few South Africans and a good many French. The Scots arrived from scuba diving and scallop harvesting and there also a lot of the British Navy people. Unfortunately, we made the mistake of giving them the top jobs and they were useless. They could swim very fast and had a very good manner, but few knew anything about diving. Later on, a new type came in from the marine salvage team and they were excellent. But the first ones were combat divers – more interested in spearing fish.' Keith said most of the divers knew each other. 'But the American companies had started to come and a British diver had no chance of getting in there. I also ran an Italian diving company and the first deep dive was an experience. Not one of them could speak English. We tended to recruit Brits and then the French started coming.' Michel said he hated diving personally, because he knew the risks. 'I was a technician and I knew how bad the equipment was. I also knew how dodgy the situations were in which we were asking people to operate. But there were plenty young people ready to do it.'

From the earliest days, these daring men were regarded as being among the highest earners in the North Sea, largely because of the risks involved. One salary level quoted for saturation divers was as high as £1,000 a week; another was £2,000 for six months' work. The former divers had differing views on the amount of wages being paid. 'To be honest, the money wasn't that good,' Michel said. 'KD Marine were paying a fixed salary between $400 and $800 US/Canadian a month in 1973–74. That was nothing like the £1,000 announced in the press.' There was also a differential between the various systems, as Keith pointed out. 'There was more money in saturation. Funnily enough, people doing saturation diving today are now in their forties and up to their fifties. You can't get them out, because of the money.' Jim Limbrick estimated that a good saturation diver could earn up to £40,000 for six months' work a year, depending on how deep he was prepared to work. Michel worked on the American barge, *Viking Piper*. 'I knew they were getting $10,000 a month and we were getting a maximum at the time of $2,000.'

The risks involved in the two systems were also different, according to Michel. 'We knew the risk of bounce dives, which involved pressurising people rapidly inside a bell while they were lowered to their working site. Then the bell door was opened as the pressure inside and out equalised. But

THE OILMEN

some of the doors were dodgy. Ours was adapted from a submarine door and we had a lot of near misses. The diver had a maximum of 25 minutes to do a task which might not always have been very well described. He never knew what he would find – or whether he would be able to see anything. Then it was back to the bell to start decompression right away.' The Frenchman said that generally diving in the 1970s lagged behind other aspects of the industry. 'We were doing our best, but the operators didn't have a clue about drilling in deep water. The subsea technology – the linchpin of all offshore activity later – hadn't then been created.' But he won't easily forget the ignorance of a number of toolpushers about how far divers can be stretched. 'I worked for KD Marine on a semi-submersible with a Shell toolpusher in charge. We lowered the bell to 600 feet to do some inspection. Then the toolpusher suggested that with a long enough umbilical, the diver could go deeper. We refused and he was furious. Our marine engineer, Alan Krasberg, who incidentally devised the standard compression and decompression diving tables, explained diplomatically that by the time we finished lowering the diver to that depth he would have less than 30 seconds to operate and we would have to recover him right away. So physically it was not really possible.'

Keith Johnson acknowledged diving was 'hairy'. 'It was a wild place to work – tough. It is relatively safe these days. You don't hear of many diving accidents. There are certainly fewer divers, but it is all so much safer with regulations and safety measures. We made all our own stuff up. The guys from the Gulf of Mexico, who had been at it longer than us, were no better. They were a wild bunch then and they still are. A lot of the old toolpushers

Plate 35.
The Leman gas field in the southern sector where diver Keith Johnston found himself trapped with a severe case of the bends. *(BP plc)*

were from the States, and it was a case of, "Get your ass over the goddam side. If you don't we'll get somebody who will."'

On some inspections, divers saw things they shouldn't have. Michel described an elderly rig, *Transworld 58*, as the 'biggest floating abortion' he had ever been on. 'It was so old, with a large metallic sling crisscrossed under the surface to keep the legs together and prevent the rig from opening up – just like the *Alexander Keilland*, the Norwegian installation that capsized in the Ekofisk field in 1980 after losing a leg. We found out that the rig was totally rotten when we had to call in a certified diver expert in underwater rust and corrosion. He was not supposed to tell us. That rig was eventually stuck in the Invergordon repair yard for years and never went back on location. They were lucky nothing ever went wrong.'

There were countless near misses as the pressures mounted offshore. Keith said he nearly died himself in the southern North Sea. 'I was trapped on the bottom in the middle of the night down at 180 feet and I got out, but I actually blew myself up and got spinal bends. I was lost and they couldn't find me at sea. I had an old Navy dry suit, with tube deflation on it. A standby fishing boat found me shouting for help on the Leman Bank. In those days we didn't have decompression chambers on the rigs: I had to be flown ashore to straighten me out. That took two to three days. That was as big an event as the latest deaths in the North Sea and it was on national news and television.'

Keith also survived a bailout problem. 'We used to dive using air from the drilling platform, which came through a hose attached to a small volume tank, down to the diver. Of course, everybody used this hose and nobody watched the connection. So when they needed the air they unplugged the hose and took it away. The diver breathing down this reservoir on the bottom underneath the Amoco A platform got nothing. That diver was me. It was only about 120 feet, but I had to ditch my gear and swim for it. The trouble was you dared not say too much because you wanted to keep the job.'

Divers often had to venture where no one else would go. Engineer Roy Wilson said that the utility shaft on Mobil's Beryl Alpha platform had flooded and the power wouldn't work. 'I was on a semi-submersible when I got an urgent call. They wanted to evacuate Alpha, bar the essential people. The flooded shaft was a job for the divers. One valve had to be closed deep inside the utility shaft – a minefield of pipes and cables and loose material and dark as anything. These guys went down through this oil-tangled black mess, found the valve and closed it. I wouldn't have even contemplated doing it. But they did it. That's the kind of guys they were – brave people. You have got to admire them. Of course, they were very well paid, until the tax laws

changed that.'

Another accident on Beryl Alpha occurred when a bell working beside the platform snarled up in a cable and dropped to the bottom with the divers in it. Other divers prepared to go to the rescue. Roy said, 'Sadly, I can recall a big argument over liability. The divers were mad and eventually brought the bell up without transferring the men to the rescue bell – because of liability. As the bell came off the door blew off. It was all put down to a fault in their own diving bell.'

At the beginning there were no laws or regulations suitable for protecting divers. Keith Johnson said BP put him in touch with John Prescott, who was with the Board of Trade at that time. 'He asked me in 1971 or 1972 to write some safety regulations which became the first-ever diving code of practice in the world.' An offshore diving inspectorate under the Department of Trade came into being in 1974, while the actual diving regulations became legally enforceable in January 1975. But the diving companies sometimes ducked even the informal commonsense rules. Alan Higgins said that if someone was needed to work extra hours, then they just got them to work. 'That was until we found about it and started clamping down. There were perfectly good medical reasons for not working long hours. Their supervisors, Australians and New Zealanders, were used to having their own way and not used to an OIM on a platform. We had a "stand up and drag out" many a time. They were the rogue element, the most difficult to control. But there were divers getting injured and killed, plus there were the long-term effects of working at many hundreds of feet and living on helium.'

For the period from 1971 to 1978 there appeared to be nothing but bad news about diving – forty-two died in the North Sea during those years, ten of them in 1974 alone. In 1975, the year of first oil, six were killed, and a year later the number was nine. From 1978 to 1983 the figure dropped to six and thereafter the fatalities continued to fall – but an important factor was the introduction of remote-control vehicles and the drive to cut costs after a harsh recession in the industry. The lists for the 1970s do not include divers who were severely injured and had to leave the industry; that number is unknown. Commander Jack Warner, Chief Diving Inspector, pointed out later that the totals were insignificant against the annual toll in road accidents or just one plane crash, but he added, 'They were too high for all involved in the diving industry. Looking back, many were easily avoidable.'

One such avoidable fatality out of that tragic roll of divers occurred on *SEDCO F* during decompression. Michel Euillet said, 'This man had a slight cold before he went diving. During decompression he complained of a pain in his chest and developed a pneumothorax. *SEDCO* had a doctor on board, but he refused to get into the chamber. Other people went in and diagnosed the trouble. But the doctor refused to acknowledge this and told them to decompress him and give him some painkillers. Unfortunately, a painkiller hides what is really happening. The diver was still suffering so they stopped decompression again. The doctor telexed onshore and they backed him in his decision. They went ahead with the decompression and killed the diver. It would have been perfectly simple to perforate his lung with a needle to relieve the pressure. That chap was only in his twenties. There was a big scandal and a court case. The doctor was banned. He knew nothing about diving. I honestly feel, however, that we, the diving team, killed that man because we never went against medical advice even though we knew better.'

The former diver said that when the large companies such as Shell started looking into the accident situation, they realised they would have to start reworking their book on safety. 'They knew nothing about diving – none of their barges or rigs had been designed for diving in deep water. Then Shell realised that, as far as public relations were concerned, it would be better to start treating not only divers, but all offshore workers, as human beings and not as expendable. They had not been that hard, just not conscious of the risk.' It also became apparent that the whole underwater sector was changing as technology became ever more sophisticated. Keith had seen the future in 1968, when Comex launched the first stainless steel diving bell. 'People thought it was something meant for the moon, but we actually ended up using it on *Sea Quest*. I suppose the real changes began in the 1970s, when there was a lot of experimental diving going on with companies racing to work out the best decompression tables to give you longer bottom time. There were millions of pounds invested – Comex in France against the Americans. Comex reached over 2,000 feet which meant breathing mixtures of less than 1 per cent oxygen. We won't know the long-term effects until the guys who did the first saturation dives die. Up to about 1983, it was mostly diving, and then remotely operated vehicles started to come in. But everything to do with marine technology was developing so fast. You would be sitting offshore and you would think of something. Could we make it – let's try it out. And that was how it would be.' Another innovation was in underwater television. Keith worked on *Sea Quest* with the first system in the North Sea. 'It was a fixed camera with a wide-angled lens and it could see the BOP. But it was always going wrong. It was always getting water inside and it hadn't proper cables, so we made our own.'

With qualifications in electronics, Michel was heavily involved in experimental technology. In 1972 he installed some of the very first guideline underwater television. 'I was dealing with the values of underwater commu-

Plate 36.
Over the side – a manned submarine heads for the depths guided by a diver on a contract job carried out by Intersub.
(Dick Winchester)

nications – all that was very new. To give you an idea, every two or three hours of diving with television required about 3 or 4 hours' maintenance. We would spend our time redoing underwater splices, inventing the technique as we went along, hoping no one would notice. Our learning curve was lagging behind.' Keith couldn't remember ever being able to talk to a diver in the water. 'We communicated by tugging his lifeline. With television now, they can watch it in the company office onshore. At that time, they relied on us to calculate something on the back of a fag packet.' The system Michel had installed was operated by remote control from the drill deck on guidelines and the camera could pan and tilt.

By then, there was another subsea intruder: the manned submersible developed from the famous World War Two X-craft and later used in oceanographic research. The North Sea oil industry recognised the value of these highly manoeuvrable, independent, safer vehicles for many of the underwater tasks. The first two subs were the *Haifa*, a sphere with small windows, and the *Perry*, a cylinder with a big front window, but both were fairly basic. Keith Johnson is somewhat scathing about the subs. 'They weren't a big success because they cost so much to support. You needed special ships and they were limited in what they could do. They tried to use diver lockout – divers going out of these submarines – but they hadn't developed the breathing apparatus to conserve the gas to give them the time out of the subs. The subs couldn't carry a lot of gas. They did develop what they called gas reclaims, which meant you used less gas.'

Keith's view is not shared by Dick Winchester, who enjoyed a career first as a manned sub pilot, then as offshore manager and latterly as operations director of Intersub, which eventually changed to remote vehicles. 'Even now,' he says, ' I get people who say, "God, I wish we had a manned sub out on this job, it would have been so much easier. The fact is that in the 1970s we did manage to injure and kill a few divers, whereas the subs were inherently incredibly safe.' Dick went into the underwater business in 1974 from the Fleet Air Arm. He saw his first-ever manned submersible, the *Perry*, in Florida. It had very basic manipulators and had to be fitted with equipment suitable for the North Sea. 'During the first year we didn't have any acoustic navigation, a very new technology, and it was a helluva steep learning curve. All the machines were run on standard heavy-duty maritime batteries. The longest dive I ever did was about 14 hours.'

During the peak construction period from 1973 to 1980, there were five manned submersible companies working all over the North Sea. There were two classes: the operations machines, with internal cameras and a crew of two or three to do surveys or inspections and handle material. The other was the diver lockout, crewed by the pilot, observer and diving supervisor and the lockout with two divers: 'The great advantage was we could see what the diver was up to and he felt very safe.' Once diving support boats arrived in the 1970s, however, that was the end of the divers' lockout from a manned submersible.

Dick never felt in danger. 'We never had a serious incident that turned out to be life threatening. We had faith in our hardware, everybody was extremely well trained and knew what to do. Operations were relatively smooth, but we did have a few interesting moments. I worked a lot on Frigg, on Statoil, doing installation work off Brent, and for Mobil.' On one occasion Dick took a submersible below one of the Statfiord platforms where it was thought pieces of scaffolding were still clinging. 'We were halfway across under the platform and this guy with me said, "How much does this bloody platform weigh?" I said, "About half a million tonnes. Wouldn't it be a bastard if it sank just at this moment?"'

The value of the manned submersible in very difficult situations was illustrated in another of Dick's experiences. 'Beryl Alpha was connected by a

THE OILMEN

Plate 37.
Pilot Ian Dempster of Intersub prepares the manned underwater vessel for a dive. *(Dick Winchester)*

pipeline to a single buoy, essentially a storage tower with a hose for pumping to a tanker. One night in December, the guys on the SBM heard a noise and somebody stuck his head out of the window. "Is Beryl Alpha supposed to move?" Somebody said, "No." "It must be us then." It had broken loose. Eventually it was towed to Stavanger where we inspected the base to determine if it had been damaged.' Dick said the hydraulic lines were in an absolute mess. 'So we had a sub which sat mid-water delicately undoing all knots in the hoses and separating everything, using good old-fashioned manipulators. An American electronics engineer was watching on the TV monitor. As the pilot I followed his instructions. I would defy anybody to do that now even with the most sophisticated of ROVs.' (remotely operated vehicles).

By 1982 industry trends were dictating change; the manned subs were being overtaken by remote vessels. 'Stolt Nielsen, as it was then, and ourselves were probably among the first companies to bring in ROVs. The reason was the change from development to production. We had a contract with Shell to inspect their FLAGS line, and it was very obvious that trying to do this robotically was a much more sensible solution. Factors like that brought the changeover.'

Like Dick, Derek Stewart from Aberdeen had no prior experience of diving. His route was through his onshore job as a television repairman. 'Comex Diving in 1975 were looking for a television engineer to work offshore and I got an interview with the chief of the electronics department. He didn't really know any more about the work than anybody else. About a day later I got a call. "Well, are you coming or not?" They wanted me for

Plate 38.
A Smit Lloyd vessel approaches Mobil's Beryl Alpha – the platform's breakaway storage unit was the centre of a delicate manipulation by Dick Winchester's manned sub. *(James Fitzpatrick)*

pan and tilt wellhead television systems to examine the BOP on the seabed. The system was always liable to be crushed. I would get a phone call in the middle of the night to be at the heliport, go out and get the system working again. Before that, they used divers, which must have cost a fortune. The first time I was on an exploration rig, the *Transworld 58*, the location was too deep for divers – a thousand feet. They were using the pan and tilt and the driller said to me, "To think we have been paying divers all these years." That was prior to the arrival of rovs.'

Sophisticated custom-built vessels designed for diving operations were starting to appear, equipped with the latest in diving gear, bells and decompression chambers. Keith worked on the first of these, a BP diving ship, the *Arctic Surveyor*. 'It had dynamic positioning, which they didn't believe would work here. But once its value to the industry was realised, the development moved so fast – they couldn't keep up.' Dick Winchester described some of the first dynamic positioning ships as 'iffy'. 'They were just unreliable. If you

THE OILMEN

Plate 39.
Diving technician Derek Stewart supervises the launch of the tiny automated machine that revolutionised the diving industry.

got a drive off [told to move away] and you had a couple of guys down in a bell, tough. This happened quite regularly and eventually they got it right. Unfortunately, they lost a few divers in the process.'

Derek, who had been working in Africa, saw the massive diving vessel, the *Uncle John*, being built in Oslo in 1972 for Comex. As diving technician, he was responsible for all the electrical work. The *Uncle John* was state of the art, built purely for diving, and it is still working in Mexico. 'I was on it on Brent for two and a half years. All the pipelines and tie-ins associated were done by divers using hyperbaric welding. They went down in a hyperbaric habitat, as big as a room. It was placed over the pipe and the water pushed out. The diver was in a bell, which was locked on to this habitat. Then the welders and divers brought the pipes together.'

Michel worked in Labrador in Canada in 1975, but when he returned to the UKCS (UK Continental Shelf) a year later, he found, like Keith Johnston, that everything had been changing fast. 'After 1975, I wasn't supposed to dive any more because my certificate wasn't operative. I was happy to work on top as an engineer. My only part in diving was to try to innovate with the gadgetry and systems we installed in the bell. It was saturation diving by then and I designed quite a few things, which I should have patented.' When he first came to the North Sea, Michel and his colleagues had urged their company to consider some form of remote submarine vehicle. 'It seemed to me totally idiotic to put people into the water when you could do it by

Plate 40.
The BP diving support vessel *Fasqadair* in the Forties Field. *(BP plc)*

submarine remote control. But they refused. Now, you can do a lot on the Moon by remote control without sending an astronaut and it costs a great deal less. If you can do it there you can do it on the seabed.' And that is what ultimately happened. Keith first used the now ubiquitous little vehicles in 1973, during the celebrated operation to rescue the crew of *Pisces*, the British Navy submarine stranded on the seabed in 1,500 feet of water. 'That was my first inkling of the potential of ROVs. I realised, "Oops, this is the future".'

The future arrived for Derek one day in 1979, when he was told to go to Vancouver to help build a ROV – the first for the company. The earliest had been built for the American Navy, the year before. 'It was a very simple machine; the motors were modified electric drills and it was all built with

Plate 41.
The massive Comex custom built semi-submersible diving ship, *Uncle John* – the state of the art diving vessel of the day.

stuff you could buy off the shelf. They were so small, only about 250 kilos, with air-filled tethers, and you could lift them over the side. Today's ROVs are basically the same, although better equipped, but they still need cables. Autonomous vehicles are still not successful because they need batteries and you can't get them back when the power runs out.'

Mike Waller laughed when he described his first experience with an auv. 'The first I ever saw was a *Scorpio*. It was a great day on *Stadrill* when this thing came – a free-swimming machine. It was launched over the side, took off and we never saw it again. Another time we had this big underwater manifold and we were worried about burying it when we were drilling. So we pumped the cuttings down a hose, which we used to inspect for wear and tear with an ROV. There was bad weather and *Stadrill*'s four big thrusters were engaged. They just blew the ROV to bits – all £750,000 worth.'

At the very start of the remote control revolution, Michel worked on a concrete platform, halfway between Norway and St Fergus, which was used to boost the flow of gas from a field on the Norwegian side to the Buchan

Plate 42.

The remotely operated vehicle – the ROV.

terminal. 'This American-made machine looked like a small orange ball. It was a fantastic gadget; it could run in and out of the structure, but it was a very difficult skill to operate it. Among the crew we had a diver who had a pet hate about remote control. Every time the machine came near him he was always screaming – we never worked out why.' Derek said they didn't need divers until things went wrong and they had to go down and fix the ROVs. 'Divers were hostile and to a large extent they were right, although we still need both.' But he admitted companies were now loathe to put divers down in the North Sea, especially in Norway. 'A lot of divers were lost; when I started about thirty had been killed. The point is that everything on the seabed is capable of intervention by a ROV and the new machines are becoming more sophisticated and computer controlled. The manipulators are so fine you can even pick up an egg. These machines are built in Aberdeen on a big scale by several companies.'

The long-term effects on the human body of repeated deep saturation dives are the subject of on-going research. The most common medically recognised conditions, which do not appear until some time after divers have ceased to work, are the crippling bone disease of osteonecrosis and the loss of hearing. Studies are also continuing into neurological damage, involving memory loss for example, as a result of decompression accidents. Michel Euillet has taken part in a prolonged long bone X-ray study since he retired. ' I am also slightly deaf and have lost some tone differentiation, apparently due to helium gas.' Mike Waller believes some divers dived too often. 'I remember one young navy guy who had obviously pitched right into the saturation diving because of the big money. I met him years later at the

THE OILMEN

Plate 43.
The next generation of remote underwater inspection craft – a free swimming autonomous vehicle – but its independence from the mother ship can cause problems. *(Subsea 7)*

Turriff show. He could hardly walk – crippled by diving. Bone necrosis. He said he hadn't known about it at the time. "I just chased the money, but money is not everything. I wish I hadn't chased it so much."'

When the machines had begun to take over the number of divers dropped dramatically. At the peak era, at the end of the 1970s, nearly 2,000 divers were recorded on offshore installations and rigs, hundreds more on support vessels. Those on pipe-laying barges were not listed. The North Sea also had the largest representation in the global diving business. Comex Engineering, Wharton Williams Taylor and Subsea Offshore, all based in Aberdeen, during the boom periods, accounted for 70 per cent of an industry worth $1 billion worldwide. Since 1986 diving has declined and today there are fewer than 200 divers on the UKCS. Keith Johnson now runs a diving company, Maris, with contracts in all the major offshore oil provinces. 'Nobody ever dreamt it would get to where it is now. It was too good to be true – a case of "get it while you can and just ride the gravy train".'

The epitaph for these extraordinary individuals is left to one of their own, Jim Limbrick. 'Divers are loud and they brag and they get drunk and they fight and bull a lot and they laugh a lot and try to be one of the crowd, but deep inside they are mostly solitary individuals and many of them actually enjoy and prefer their own company. This stems, as is obvious, from the fact that their underwater life has a lot to do with self-preservation in a lonely world where, even though there is someone on the end of a phone, they may just as well be a million miles away.'

The Supporting Cast [6]

Behind the headline glamour of the drilling rigs and the parade of dramatic oil and gas strikes since that very first foray across the 56th parallel, there has always been a wider and disparate supporting cast of thousands who have been instrumental in transporting, maintaining, feeding and protecting the front-line troops. A Government survey in 1980 revealed that 29 per cent of the workforce were employed on offshore construction; 32 per cent on production and maintenance; a further 16 per cent were engaged in drilling activities; 15 per cent were involved in domestic work, wireless operations and as ships' crew; 7 per cent as helicopter pilots and on pipe-laying barges, and less than 1 per cent working in the diving sector.

That supporting cast is an essential and integral part of the North Sea oil story. Its members are to be found in the advance scouts of aerial and marine geologists and surveyors; operating helicopters and the busy fleets of freight line supply boats to the scattered colonies of rigs and platforms; on the vital standby and rescue vessels which circle the installations monotonously; and in the run up to first oil, on the giant barges which disgorged at a remarkable rate, the hundreds of miles of pipe that now form the grids of subsea networks transporting the oil and gas to the terminal landfalls. They are also found, of course, on the production platforms themselves – the contractors with their busy armies of engineers and technicians, and service, maintenance, medical, communications, housekeeping and catering staff. All are companions in adversity, with the rig and operator crews, fighting the hostile weather, the unpredictable seas, and all are constituents of the same unique and artificial industrial community. Forming the supportive base of this pyramid is the onshore network of company offices, bases, terminals, plants, yards, workshops and clusters of ancillary industries which have sprung up around Scotland and elsewhere in the UK in the incredibly short span of thirty-five years.

If, as Commander Warner claimed, that without divers there would be no oil, it is equally true that without these many other players there would be no

modern industry. In the run up to production, with the workforce beginning to build up for the construction phase in the mid 1970s, there were 9,200 people employed on the installations and rigs offshore while there was an estimated 2,000 or 3,000 more involved in the transport, support and service sector. At the peak of the production phase in the 1980s, total employment in the oil and gas industry in the North-east of Scotland was 52,000, with more than 21,000 engaged offshore. Among them were the pilots and crews of the helicopters.

The airborne ferry

On 27 July 1967, a Bristow Whirlwind Helicopter made the short trip from Dyce Airport, Aberdeen, to the RAF base at Kinloss, in Moray. There it picked up a crew of oil workers and took them to the drill ship *Glomar IV*, which was exploring for Hamilton Brothers in the Moray Firth. That historic journey was the inaugural offshore transport flight to the northern North Sea – the prelude to many thousands of airlifts of men and materials by the helicopter fleets which undoubtedly made possible the accelerated growth of the oil and gas industry. The synergy of the helicopter and the hydrocarbon developments offshore is one of the happy coincidences of the emergent new technologies of the late twentieth century.

The two found common cause at precisely the right time. The oil companies were perfecting the means to explore and exploit the unknown marine environment and the aircraft had achieved astonishing versatility of performance in the Vietnam War, militarily as a gunship, but significantly also as a willing transport workhorse to areas inaccessible to fixed wing aircraft. Previously, the early helicopters had been used extensively for commercial and agricultural purposes in the United States. Then, some years before the southern North Sea operations began in the 1960s, the aircraft found their most important civilian transport role in the oil fields of the Persian Gulf. Bob Balls, who had learned his trade in the navy, worked in the Middle East. 'We were flying for Bristow, but we were contracted to Pan Am, an offshoot of Amoco. Then we started in the southern North Sea from Tetney Lock at Cleethorpes in Lincolnshire, in 1965.' The oil companies quickly realised the helicopter was the key to harvesting offshore oil by providing a reliable, swift airborne and cost-effective conveyor belt to the rigs and platforms.

Bob Balls' experiences are a fascinating chronology of the development of the now familiar mode of transport that virtually opened up the North Sea.

Plate 44.

Oilmen return from a trip to the southern gas fields to the small Bristow base at Tetney Lock at Cleethorpes in Lincolnshire, in the 1960s. *(BP plc)*

'At Tetney, we started with Whirlwinds – the smaller helicopters. Then shortly afterwards, we got the Wessex. The Whirlwind was a single engine on floats. It carried eight passengers while the Wessex was twin-engined and held sixteen. By then I was at Yarmouth, where BP had a base. Working from a small place like Tetney in the beginning you knew most of the passengers. In the Wessex you had to have a passenger in the back connected to the pilot by headset. He would let you know if there was anything happening behind you.' Bristow had begun to operate out of Aberdeen, but Bob worked abroad for a year. 'I came back to the North Sea in 1969 and it really kicked off for Bristows, although British Airways were there before that. In Aberdeen we started off with five pilots – two crews and a spare pilot. I helped with the start- up there as a training captain.' British European Airways had already established a base by August 1967 and their Sikorsky 61N – the more familiar workhorse of the North Sea – made its first flight offshore. The following year BEA carried out the first casualty airlift and the first evacuation in the Northern seas – both involving Shell's *Staflo*.

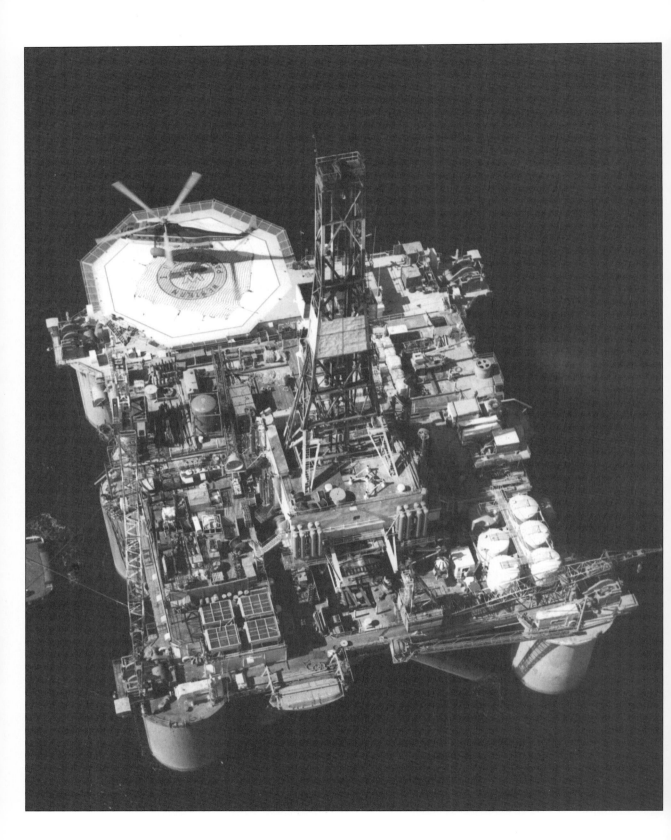

THE OILMEN

In a previous chapter Swede Lingard referred to Bob's bravery. This stemmed from his daring evacuation of a rig crew, the first UK operation of its kind from an oil installation. In March 1968, the semi-submersible, contracted by Burmah Oil to drill 125 miles off Scarborough on the Yorkshire coast, was pounded virtually in two by 50-foot waves in 90-mile-an-hour gales. There were forty-five men on board. Bob Balls was the only pilot near the scene. He recalls that night as 'the hairiest' in his career. 'You have to remember we had evacuated the *Sea Quest* several times already, but we knew it wasn't going to sink, whereas the *Ocean Prince* was moving in two ways and breaking up. I was sure the bloody thing was going down. Landing wasn't what was worrying me. Don't forget there were forty-five men, so I had to get one lot off, get them to another rig 16 to 18 miles away, go back for the others, land again and get them off. I made three trips all told. The worry was that you wouldn't get back before it sank – which it did about an hour or two after the last trip.' For that feat, Bob was invested as an MBE. The citation said, 'but for his initiative, bravery and splendid airmanship, the members of the *Ocean Prince* crew would have probably lost their lives'.

By 1972, the helicopter operation was expanding rapidly at Dyce. BEA Helicopters opened a purpose-built complex and transferred their maintenance operation from Heathrow to Aberdeen. Bristow brought in their first S-61N, which was capable of carrying nineteen passengers. Flying out of the growing heliport at Dyce was a real contrast to the easy-going days at Tetney Lock. 'Those were our own bases at Tetney and Yarmouth. At Aberdeen you suddenly had to abide by airport rules – wait in lines for take off – that was quite different. We had to lease a hangar at first until our own set up was built. They brought in more of the S-61s through the 1970s and then in the 1980s you had the Aerospaciale – they called it the Tiger – which was faster. There was the S-76, which was slightly smaller but faster, and mainly for offshore inter-rig work you had the Bell 212.' There were also restrictions introduced by the Civil Aviation Authority (CAA) on weather limits. More sophisticated instrumentation came with the introduction of the larger machines. Each of the giant Sikorskys cost, at that time, around £800,000. For every hour it was in the air – again at that time – it burned up an average of £300 to £400 of fuel. From both companies, a total of twenty-eight pilots flew out to the rigs, which had been increasing in numbers. The Bristow operation then doubled within six months with three helicopters in use, while BEA employed six machines. Between them they flew more than 4,200 flights in 1973.

The journeys had settled into a pattern. Helicopter pilots like Bob and a

Plate 45.
The Sikorsky 61N which became the familiar means of transport over the greater distances, loading on the rig *Western Pacesetter*.
(Malcolm Pendrill Ltd)

THE OILMEN

Plate 46.
Winching material aboard by helicopter from a supply boat which has had to lie off the installation. *(CHC Scotia Ltd)*

fellow airman Richard Enoch are pragmatic about what they have to do. 'It does need considerable skill, training and commitment to actually do it,' said Richard, 'but ultimately it is an airborne bus.' Bob was philosophical. 'It was never boring. I had quite a lot of variety and the weather is always different. I would say it was a thinking man's job.'

And there is always plenty for the pilots to think about. The oilmen's 'bus' ventures further than any other commercial helicopters and in the worst of weathers. Bob described what was involved in pinpointing a solitary target in the vastness of the waters. 'It was like a postage stamp in the middle of the North Sea. Finding it was tricky sometimes because navigation was on Decca

Plate 47.
The divisive lives of the offshore workforce – the start of another two-week stint. *(CHC Scotia Ltd)*

radar in those days. The further you went out the less accurate the radar became at times. I have missed a rig because of that and had to come back. But also you had to consider what you would do in the event of engine failure. You can fly on one engine, cruising, but you might not be able to land on one engine – not on a rig, because it takes more power to hover than a running landing. So if you had an engine failure, in reality, you should divert and come back to base. The worst place right in the middle of the North Sea, I can tell you, is 180 miles from Scarborough, 180 miles from Aberdeen and 180 miles from Esbjerg in Denmark. Normally you would say OK, I will go for 200 miles and if I miss the rig, do another 100 miles and hit the coast of Norway as a diversion.'

Former RAF radio officer Mike Jennings was the helicopter flight information officer on Piper Alpha. 'There was probably a lot more helicopter traffic on it then than there is now on Piper Bravo, and there was a lot of responsibility because there wasn't the air traffic coverage from Aberdeen they have now. So the FIO acted more like a traffic information centre to the helicopters.' It was an extremely busy operation. 'We did quite a bit of overtime. There were probably about 100 flights a day, back and forth, and they asked us to keep an eye on them in case they couldn't raise Aberdeen. They had a radio blind spot somewhere and sometimes we had to stay on shift up until nine o' clock at night.'

Later in the decade, helicopters based at Sumburgh on mainland Shetland and on the most northerly isle of Unst serviced the drilling installations in the developing Eastern basin. For the North-east and north communities, a welcome by-product of the helicopters was the introduction of a search-and-

Plate 48.
A Bristow's SA3321 Super Puma, adapted for North Sea flying with sponson mounted flotation devices which were developed by the company's own design office.
(Bristow Helicopters)

rescue service by BEA, complementing the traditional land and sea support provided by mountain rescue teams and by RNLI vessels. Aircraft from RAF bases have widened the scope of the service, while the commercial helicopters are now controlled by HM Coastguard.

In the 1970s and 1980s, as the offshore operations intensified, the oil companies, with their reputation for wanting things done yesterday, increased the pressures on the helicopter firms to tailor their operations to their 24-hour schedules, flying at night and in all weathers. Over the next four years, a whole new ancillary industry was built up as the new heliport underwent a colossal expansion; Bristows now had 370 staff including 120 pilots, flying eighteen S-61Ns and five S-58Ts; British Airways had nineteen S-61Ns, 130 staff in Aberdeen and seventy in Sumburgh, including sixty pilots. The total number of trips by the two companies in 1973 had risen to 35,000. Meanwhile, a third company, North Scottish, had set up at Longside, near Peterhead with seven helicopters and a staff of sixty. The helicopter companies were being pushed to fly in uncertain weather but, according to Bob, sometimes the operators didn't want to pay for it. 'I remember I had done a crew change, a day's work – and I went to the bar for a drink. This oil chap said, "You can't drink, you are a pilot." I said, "Bugger off. I have done my day's work. I am not on duty anymore." You had to tell people that you couldn't be on 24-hour call when there was only one of you. That was the result of penny pinching in the early days.'

In 1977, Bristow were hit by what proved to be a damaging strike by fifty of their pilots, later described in the company's history, *Leading from the Front*,* as 'a watershed'. The dispute began initially over the refusal of one pilot to transfer to Nigeria but widened into a confrontation over pay and conditions. Bob was one of seventeen pilots who refused to strike. 'I walked through the picket line. I didn't belong to any union and I didn't agree with what they were doing anyway. British Airways were totally unionised and their pilots came out in sympathy. So you had the stupid situation that Bristow were flying and BA weren't.' All the protagonists involved were later roundly censured by a public inquiry.

In the aftermath, a young graduate of Bristow's Training School arrived at Dyce. Twenty-four-year-old Richard Enoch, from Torbay in Devon, had been a private pilot who decided that commercial flying would make a good career. 'I had actually got married during the course and we moved up here.' Richard, now operations manager at Dyce, remembers his first flight with a training captain. 'Training now is considerably more comprehensive, but you would practise landing on rig helidecks. The first one I did, on Forties, when I looked down I thought, "My goodness, that is too small to land on," but we landed on it anyway.'

The weather was another revelation for the young pilot, although he had been aware of how bad it could be. 'The weather was and remains the biggest difficulty in the North Sea. It is so unpredictable. The wind gives us problems in terms of sea state. By that I mean if you ever have to end up in the sea the rescue vessels can't operate. A helicopter actually performs better in strong winds, until you start getting turbulence. On platforms there are so many obstructions you can get quite a bit of turbulence.' Then there is the problem of visibility. 'We have had pilots out working the top end of an oil field and they would radio, "The weather is getting a bit bad out here, what is it like with you?" You would say, "It's fine – come on back." In the ten minutes they took to come back you would be in fog and they would have to go on somewhere else.'

Richard considers that because of having to negotiate a number of hazards such as sea fog, the North Sea pilots are more skilful than their onshore counterparts. 'In 1980, we used to do an inter-rig shuttle run with the Bell 212. They would want to move 100 people from this platform to

---

* Healey, A., *Leading from the Front: Bristow Helicopters – The First 50 Years* (Stroud, 2003).

THE OILMEN

that flotel [accommodation vessel] – ten people to a helicopter. So there would be two aircraft, one up and one down.' In 1990, there were five oil field helicopters, with seating capacity of between four and thirteen, based in the UK sector. 'At that time the flying we were doing was probably unique to the North Sea, and more skilful. Because we are flying over water, we have to monitor our fuel rather more carefully than an average helicopter pilot. If they get it wrong they can land in a field; if we get it wrong we end up in the water.' Richard has never had to ditch in the sea. The pilots train for ditching, using a helicopter simulator, an emergency evacuation trainer and a helicopter underwater escape trainer. They also work with life rafts and practise escaping from an actual aircraft.

At the beginning, the financial rewards for the North Sea pilots didn't match their skills. 'I came straight from college and the wages then were definitely lower than commercial fixed-wing pilots. But over the past four years the salaries have improved markedly. We are not on a par with 747 captains, but our money is equal to wages of fixed-wing pilots and maybe better than some.'

Richard worked out of Shetland earlier in his career, doing day trips in an operation they called 'the Sumburgh Double Shuffle'. 'We flew from Aberdeen to the Beryl field, back with the crew to Sumburgh to meet the fixed wing flight, swapped passengers, took them to the Beryl and then came back round to Aberdeen. Offshore you actually lived in a hangar on a flotel and I did that for five years.' For two of the years he was based in Shetland, Richard was also responsible for search and rescue flights.

The traffic movements out of the world's biggest and busiest heliport is the weather vane of the industry – the fluctuations in the volume of daily flights offshore are a potent indication of the health of the oil and gas world. Since 1997, helicopter passenger figures have dipped from 479,100 to a low of 385,900 during the 1999 slump, only climbing back to 438,00 at the end of 2002. The decline in business has prompted a flurry of redundancies over the years over all the companies. Since the frantic days of the 1970s, the helicopter companies at Dyce have undergone substantial regrouping operations – firms going out of business, others changing hands or merging, and some reforming. Now there are three main companies based at the heliport – Bristow, with forty-five aircraft and a 51 per cent share of the market, Bond, with six helicopters, and British International with thirty-six. The last, originally a British Airways offshoot, is now owned by the Canadian CHC Helicopter Corporation, the world's largest helicopter company. Bond Offshore Helicopters have returned to the North Sea and won the biggest-ever contract from BP against fierce competition from the other companies.

THE OILMEN

Plate 49.
The distinctive scarlet livery of the Bond Offshore Helicopter company, which had sold out its business at one point but is now back in the oil industry aerial ferry business. *(Bond Offshore Helicopters)*

Flying has now changed immeasurably, since Bob Balls' pioneering journeys out over the southern North Sea, while pre-flight safety instructions for passengers are also very different from those early days. 'Then, the helicopter carried a dinghy and all the passengers had Mae West life jackets. You had to go out to brief them. By the time you got to the 1980s, the passengers were shown a video, and very often you would have a cabin attendant. You never had any contact with the passengers anymore and you didn't know them.' Just after Richard arrived in Aberdeen, immersion suits were being introduced and now lightweight survival clothing is worn. Flying technology is also constantly evolving, with new aircraft, changes in navigation equipment, lifejackets and beacons, and in onboard systems to monitor the health of the helicopters. There are also flight-time limitation schemes imposed by the Civil Aviation Authority and control over the hours worked. But despite all the modern safety provisions, the helicopter journey is still regarded as one of the most vulnerable elements of a risky industry. Since 1967 there have been six tragedies – the worst in 1986 when forty-five people died in a Chinook helicopter in Shetland. Another major accident was the loss of eleven men in 1992, when a Super Puma ditched and overturned near the Cormorant Alpha platform. In 2002, eleven offshore workers died when their Sikorsky S76 plunged into the sea in the Leman gas field in the southern sector. There have also been numerous near misses and many oilmen have a helicopter 'experience' to recount.

Plate 50.
Helicopters on parade – the impressive line up of the North Sea's airborne ferries on the tarmac at the Dyce Heliport, Aberdeen.
(James Fitzpatrick)

Graeme Paterson, a long-serving offshore painter is one. 'I was working on Conoco's Hutton tension leg platform (TLP) over Christmas and I was told I could go home on a chopper coming on Boxing Day to pick up a compassionate leave. The weather was so diabolical we had to hold on to a rope tied from the handrail to the chopper. The pilot had already done a few drops, and because of the weather, we had to stop on a semi to refuel. Well, that was the worst experience I've ever had on a chopper. I thought I was going to die. This thing comes into the semi, which had a double derrick, and the wind is so horrendous the chopper ends up all over the helideck on its tail. You're sliding about trying to see where the standby boat is and it's actually sitting on its tail, and you are thinking, "If we go into the drink, forget it." Then the wind flips the chopper back up and the pilot manages to land. Everybody had to get off while it refuelled. Seventeen guys, all just pure white. Two wouldn't go back on. Of course, the helicopter should never have been flying, the weather was so bad, but that pilot did well. One of the guys reckoned that when the chopper flipped, the tail rotor was only three feet off the deck. If it had hit that would have been it.'

The supply chain

At water-level, life could sometimes be even more hazardous – especially in the precarious early supply boats. 'Waves crashing over the boat – that was

Plate 51.
A big change from the early days of the crew change operations, when there was little or no survival or safety advice offered. Offshore passengers, clad in survival suits, are given a thorough briefing by Bristow staff before embarking at the Aberdeen heliport.
(Bristow Helicopters)

very, very common. Basically, you had to do a little dance on the deck sometimes. Eighteen inches of water coming up that back end of the boat would knock you off your feet – it is unbelievable, the power of water. Two feet of water would have the containers floating around. You really had to be on your toes – it just seemed so open. I quite often got washed up the deck. You would always keep an eye on the arse end of the boat and try to judge which wave was going to cause you problems. Ninety-nine per cent of them wouldn't come on. I have seen a 10-foot wave come over the back end of a boat when I was on it. That was one of those moments you say, "Oh, oh. There is just nothing I can do here,"' says Ian Sutherland, from Portsoy in Banffshire, talking about what it was like to work as a crew member on a supply boat.

The first vessels had been steaming out of the harbours at Aberdeen, Peterhead and Lerwick since the end of the 1960s. As the fields began to open up, more than 1,200 supply boats and safety/standby vessels were based in the port of Aberdeen. By the 1990s, the traffic had swollen to 5,000 ships. Until new marine technology took over there was a motley fleet of converted trawlers, coasters and merchantmen of all ages and levels of seaworthiness,

Plate 52.
The endless supply chain – a transport vessel lies off in a heavy swell.
(Allan Wright)

crewed by the typical oil industry League of Nations. The Americans had brought tugs, but they were unsuitable for the North Sea. Gradually, tough little custom-built boats began to appear with bow thruster propellers and distinctive flat decks designed for containers of heavy engineering supplies and tanks for water, mud and cement.

Ray Craig had started in the Merchant Navy, moved to the Army and left in 1975. While waiting to study law at Aberdeen University he joined a Ben Line supply ship and within six weeks was sailing as skipper. He maintains that supply ships were more dangerous then than normal vessels. On one occasion he arrived off one of the Forties platforms. 'We are backing up to the rig and literally working the two main engine handles in the bridge. The bow thrusters are useless, not enough power. You are having to hold her and the wind is in your bow. You fall away. There is nothing you can do about it. By then, I am right underneath a platform producing a huge part of Britain's GDP. We are rolling violently. I have a totally pissed chief engineer, a young Scots lad, who is staggering up and down shouting at the top of his voice. Right above me are all these immensely heavy overhanging hoses. I put the engine full ahead and I am looking aft. I put the tiller on dead centre and I can only go out in a straight line. The drunken chief engineer I am trying to kill but cannot reach him. There's a cacophony of noise. We shoot forward

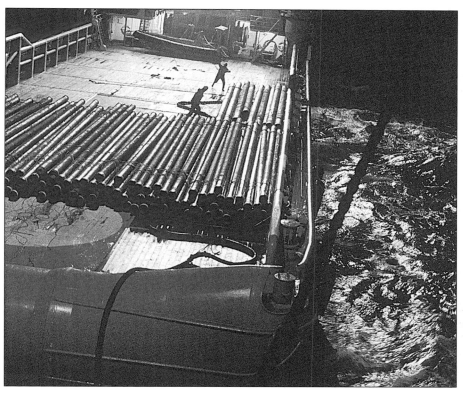

Plate 53.
A cargo of pipes as a supply boat backloads.
(Allan Wright)

and the crew are running up the deck. Over the radio the voice of the American tool pusher, "Captain, you are just a little bit fine. Can you come further out?" My hair is standing on end, but I am actually acting quite calmly. There is nothing more I can do. It is full power – and hit or miss. I say, "That's not your usual complaint – too fine." Then we come rocketing out. Of course, you can't say to yourself, "I want to go home – I want my mammy!" If anybody had seen what had happened . . .'

The supply and standby vessels transformed Aberdeen Harbour from a modest port dependent on trawl fishing and coastal traffic to a major European maritime centre. By 1977, there were no fewer than seven bases devoted exclusively to oil, operated by Shell, Amoco, Total, Chevron, Texaco – and two indigent Aberdeen companies, the Wood Group and Seaforth Maritime. An eighth base for common use was opened in 1982, followed by a ninth, leased to Salvesen Offshore. Other bases were opened at Peterhead by the Aberdeen Service Company (ASCO), at Lerwick and Montrose. There was fierce competition among the companies, who included Brown and Root, Halliburton, Maersk, Smit-Lloyd, Stena Offshore, Stolt Offshore and a Scottish company, North Star, a traditional trawl fishing company which had converted its vessels to supply boats. The company is now the supply division of the Craig Group.

Plate 54.
A Smit Lloyd supply vessel ploughs her way through waves that can sweep all before them when unfortunate crewmen forget to take precautions. (CNR)

Ian Sutherland had been told life on the supply boats was hard work, dangerous, cold and miserable and he wasn't keen to go. But in the 1980s he alternated between the Merchant Navy and the supply-vessel business, ultimately working for Wimpey Marine on anchor-handling supply ships. He loved it but admits it was risky. 'On this particular day a wave came over higher than the crash barrier. I just dived into the space for the winch. God, I don't know why no one was killed. The three of us were battered senseless. The weather wasn't really bad, I have certainly been out in a lot worse. So it was either a freak wave or we just caught it wrong.' Ian moved to the rigs and his wages jumped 50 per cent. He also worked as a crane operator on a diving support vessel.

The statutory requirement for standby rescue vessels only emerged after the *Sea Gem* disaster exposed the need to provide a 'safety net' for installation crews. Under the Emergency Procedures Regulations, each installation had to be shadowed by a dedicated standby ship operating in a zone of 5 nautical miles. The vessels were certified under a Department of Transport

Plate 55.

An early fast rescue boat on a sweep of the seas round an offshore production platform. *(BP plc)*

code of regulations drawn up in 1974 and agreed on a voluntary basis by the UK Offshore Operators' Association (UKOOA), but the scheme has been a constant source of conflict between the operating companies, the unions and the government. The operators, concerned at costs, have always pressed for self-regulation while the unions want the prescriptive measure to remain and be improved.

The original standby vessels and their crews who spent their weeks endlessly circling the installations were most definitely the poor relations of the growing North Sea fleet. Many vessels were totally unsuitable and there was low morale among the crews in a boring job. Mel Keenan, a young trade union official in Aberdeen, took it upon himself to investigate the standby business. 'The official marine surveyors were responsible for surveying the vessels and training crews. Within the 500-metre zone, the vessels were the responsibility of the OIMs. The National Union of Seamen handled the supply boat crews but nobody was interested in the standby fleet.' Mel said standards weren't what they are now. 'I had a go at one vessel, which had

THE OILMEN

Plate 56.
Contrast in provison – a modern state-of-the-art daughter rescue craft, *Delta Phantom DC* operated by North Star. *(North Star/Craig Group)*

replaced an Aberdeen ship. One crew member had been a Dundee car salesmen and never been to sea before. That kind of alerted me. The crew didn't have proper protective equipment, nor had they undergone any training. It was an Aberdeen company but the ship had come from Lowestoft and really was in quite poor shape. The Board of Trade inspectors detained the vessel and made them rectify the deficiencies.' But he said the conditions and the standard of standby vessels improved. 'Once they demonstrated their capabilities the offshore workforce grew to depend on them. They were a comfort as it were, and provided a feeling that you were being looked after.'

It needed another disaster before the authorities looked again at the capabilities of the standby rescue fleet. One of the most memorable feats of sheer courage during the terrible night of the 1998 Piper Alpha tragedy was the repeated heroic sorties by fast rescue crews to pick up survivors. But in the inquiry's 1990 report by Lord Cullen, the general standards of the ships were criticised by survivors and other witnesses. At that time, of the 187 vessels recorded, 162 were converted trawlers. There were only seven purpose-built vessels – mainly Norwegian – and eighteen part-time multi-functional ships. The report noted that 'the costs of the operations have not enjoyed a high priority in the operating budgets of the oil companies'. Lord Cullen then 'strongly urged that the standards be improved with despatch'.

Plate 57.
A supply boat lies off in the Tartan field – in the distance, circling protectively, a stand by ship.
(James Fitzpatrick)

An upgrade of the voluntary code of regulations was finally put into effect in 1991 and 104 vessels were withdrawn and 155 others improved at a cost of £200 million. Now the towering multi-functional vessels combine supply, rescue, standby and firefighting capabilities, and some even perform as mobile drilling rigs. Currently BP are working on Operation Jigsaw, which involves replacing the chartered standby rescue vessels by three giant regional support vessels, complete with two autonomous recovery rescue craft and supported by inter-rig helicopters. BP deny claims they are downgrading the safety regime, but unions have pointed out that helicopters were powerless during Piper Alpha and were unable to pick up any survivors.

The advance guard

The American geologist leaned on the rail of his frail-looking craft as it moved against the quay in the waters of Aberdeen Harbour. It was 1965. The tall Texan, inevitably chewing a long cigar, was talking guardedly about why he was in port. He pointed to a stack of crates on the decks. 'That there is dynamite and we use it to shoot explosive charges for seismological tests. We're trying to find geological formations that might hold traces of oil.' And,

THE OILMEN

Plate 58.
The formidable vessels that now perform a multitude of tasks offshore – the *Toisa Proteus* is one of three massive sister ships owned by Sea Lion Shipping. Her principle role is as a support for saturation diving and underwater construction. *(Sea Lion Shipping)*

was there oil? 'Hell, there are millions of dollars sunk in this thing. Oil companies don't spend all that damn money for nothing.' The Texan was one of the geologists and surveyors who formed the vanguard of the hunt for oil across offshore Britain, funded by the oil companies even before the sea's acreage had been carved up. Nigel Anstey maintains in the Petroleum Exploration Society anthology, *Tales From Early UK Oil Exploration 1960–1979*, that 'the success of the North Sea was the success of geophysics. Particularly in the oil province where there were no onshore leads, the best geology in the world could never have discovered the riches beneath that featureless sea.' He wrote, 'The North Sea emerged just as the seismic method was taking a great leap forward; quite fortuitously, the problem and the solution came together at exactly the right time.' He was talking about the technical advances in the 1960s and 1970s, including digital recording of seismic data, non-dynamite systems such as the air or gas guns to fire the seismic shots. The next big advance in seismic surveying was in 1975 with the introduction of 3D technology. The future now lies with the exciting advances in the use of 4D technology.

But it was all a mystery to seventeen-year-old Mike Standish when he decided to finance his university studies by signing on as a lowly crew member on a seismic survey vessel. The experience was so traumatic he swore never to work offshore again. 'My uncle was with Seismograph Services and he wangled the job for me. I was a bit young and this was 1973, when health and safety wasn't a strong culture. I didn't receive any training. I just turned up at Immingham on the Lincolnshire coast and got on board this converted Norwegian trawler, the *Kristian Tonder*, a 1,000-tonne vessel with a ramp off the back.' The crew were Norwegian and the technicians were mainly Brits, about thirty or forty people. The *Kristian Tonder* sailed north round the top of Scotland, through the Pentland Firth and into the Celtic Sea, trailing cable and firing shots. Mike's job was mainly changing the data recording tapes, old-fashioned one-inch seismic reel tapes with big aluminium covers. 'Every 30 seconds, there was a pop – a shot that is, and there were about seven shots a tape. Gas guns were the firing source and every couple of minutes I had 30 seconds to change the tape. Of course there was a problem if I missed a shot and loused it up. But they didn't do anything, it was just a missed shot. It was all two-dimensional in those days, no such thing as 3D seismic. As for safety, I think eventually they dug out some safety harnesses, but basically you worked without them. Every seventh wave – this magical seventh wave – would come right up the deck and there would be a shout; you would jump, hang on to a pole running across the back of the deck, wait for the wave to pass and drop down again.'

Mike said it was a six-week trip. 'I remember going round the Pentland Firth, which is a very busy shipping lane. It was very foggy and the radar had broken down. So we sat the cabin boy in the bows with a walkie-talkie, spotting ships. If he saw something he would radio the bridge, which struck me as being a bit dangerous, but we got round.' On one occasion, Mike found himself in real trouble. 'Below decks, we would cut open the plastic cable which was full of kerosene to make it buoyant and tear out units to be used again. Of course, kerosene spilled all over the place. As I was pulling cable, the metal bit fell across the terminals of an old car battery – I don't know why it was there. It sparked and there were a few flames. There I was ankle deep in kerosene. So I shouted, "Fire extinguisher." A guy grabbed one but it didn't work; he grabbed another, which didn't work. So he shouted to the guys on deck for another fire extinguisher. They dropped one down, but it still didn't work. Fortunately by this time the fire had gone out and it never actually reached the kerosene.'

Gradually the seismic operations in the North Sea became more sophisticated as techniques and equipment improved. 'Boats were also bigger and more luxurious, plus there was more emphasis on health and safety. Nowadays they pull multiple streamers in huge arrays and have multiple gun arrays.' Mike graduated in geology but he stayed on land with data company Seismic Services. It was bought by Schlumberger and he is now their site manager at BP's business headquarters in Sunbury.

Once oil had been found in commercial quantities, the surveyors had to

THE OILMEN

investigate the stability of the seabed sites for the platforms and map out routes for the pipelines. By the time electronics engineer Barry Matthews started in the North Sea, he already had experience with Hunting Survey in the southern sector in the early 1960s. In 1973, when the surveyors followed the drillers north, where the water was much deeper, techniques used in positioning and sub-bottom investigations had become more sophisticated. 'The information had to be more accurate and we worked subsea rather than from the surface. Usually it was OK, but later on, in the late 1970s, there was a big problem with gas bubbles, or pock marks. Basically, the gas permeates from away down and leaves a big pockmark on the seabed when it breaks through; you don't want to plant a platform there. We used shallow seismic to look at the top 50 metres of the structure to see if it would support a platform and if there was any evidence of gas bubbles. That never happened with an installation, but there were a couple of times pipelines were threatened by these gas bubbles. From the beginning all the pipes were buried, unlike the Gulf, but one of the main problems up in Shetland was that the routes actually crossed buried rock outcrops. That was when the surveying became quite intensive. We had to find a way past that. You can't exactly bend a pipeline.'

Life on the early survey boats, according to Barry, left a lot to be desired compared with the 'luxurious' custom-built vessels with single cabins and showers. 'The really early ones were converted fishing boats, which still stank of fish, and pretty dismal tramp steamers, coasters – anything they could get hold of. Later, we worked on supply boats, but there wasn't much room for people.' The biggest enemy, as always, was the weather. 'You go north and we are talking about 12 hours' steaming, which needs fairly big vessels to be able to stay out. They never considered the people. One manager told me, "We can replace people, they don't cost much, but the equipment, now, that is very difficult to replace."' Barry ended up as party chief for Brown and Root and later with Subsea, which took over the survey work. 'I became a multi-tasking person when there was a shortage of personnel and it actually saved money. In the early days the salaries were quite good, then staff wages started to go down while freelance or contract guys got a lot more.' He now works with his wife in her costume and dress hire business in Aberdeen.

The pipe layers

As the race to first oil quickened, almost in tandem with the seabed surveys, the pipelaying sector had already swung into action, digging seabed trenches

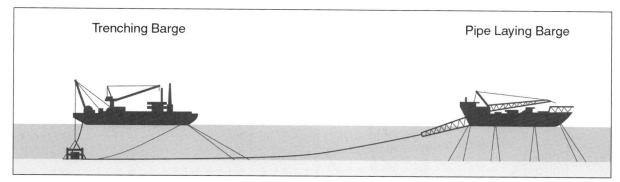

Figure 3.
One version of the pipe-laying barge systems that created the network of transport lines for gas and oil – complete with stinger and array of anchors. The vessels followed the trenching barges that gouged out the seabed troughs. *(Greybardesign)*

from the wellheads to the shore while spooling out the steel cylinders that would transport the oil and gas. Hundreds of miles of massive pipes were churning out from the mills and coating plants onshore. The pipes, in 39-foot lengths, were usually between 20 and 36 inches in diameter. To survive the seawater they were coated with a thick layer of anti-corrosive material with a layer of cement to prevent them buckling. The pipelaying specialists were the Italian and Dutch, with Americans adding their skills and expertise. The first pipelines were laid from the Ekofisk field in the Norwegian sector, one transporting oil 350 km to Teesside in the UK and the other gas, 430 km to Germany. Huge Dutch barges, *Choctaw I* and *Choctaw II* laid 1,700 metres of pipe a day at an estimated operating cost of £1 million daily.

John Carter, a master mariner from Caithness, and a land pipeline worker, Brian Porter, from Derbyshire, were among the awed oil company observers on the first barges plying to and from Forties in the 1970s. Brian, a toolmaker, had begun by installing gas pipelines onshore, but after qualifying as senior pipeline inspector, he worked in pipe mills and coating plants. Then he was sent offshore in 1973 as an inspector for BP on a lay barge. 'That was an Italian vessel, the *Castora II*, owned by Saipem, laying pipe from Cruden Bay to the BP Forties field. There were a lot of Italians obviously and their underlings were Portuguese – treated like bloody slaves. There were about 300 people, including welders, coaters, radiographers and contactors, and four inspectors – I was doing coating inspection. You went on and off by helicopter but you couldn't go off in rough seas. If the barge stopped laying, it would either anchor up or it would be towed back to Peterhead; they didn't have engines. So we would get off and the barge would go back out. We returned on a supply boat and, believe me, it could be very rough.'

Life on board the barges was limited and deafening because of the nature of the 24-hour operation. 'The accommodation wasn't bad, but it was very claustrophobic, with nothing above decks apart from pipe, the helipad and where the pipe was laid. It was very noisy, the cranes were going when the pipes were going and it was difficult to sleep. Every time the crane came back

down, the whole barge reverberated with a boom. The canteen was quite good. A lot of Italian food, but steak and eggs for breakfast, although it was probably horse. It didn't matter, you were that hungry. We also got wine and bottles of beer; Dutch lager, as weak as gnat's pee. You weren't allowed to drink on shift, though some of them did. There was almost a blind eye in those days.'

The raw 40-foot pipes arrived in Britain from steel mills in Germany, Italy or Japan to the coating plants. Brian remembers one job that involved 48,000 pipes and another for British Gas, with a throughput of 40,000. The pipes were then despatched by supply boats, 24 hours a day, to keep the lay barge going. Captain John Carter was also involved with barges laying pipe a year before Brian Porter went offshore. John is from a prominent Caithness fishing family and he had gone deep sea and qualified as a master. In 1967, he became harbourmaster at Fraserburgh; he was also lifeboat committee secretary when the lifeboat was lost in 1970. He commanded a number of North Sea supply and support vessels, notably during the Forties construction. His ship had towed one of the lay barges, the *LB Meadows*, to the BP field while the *Christola V* came out from the shore. 'We were loading from two distribution points, Invergordon or Peterhead – they were laying so quickly, there had to be two distribution points.' The pipe was coated at MK Shand, at Invergordon. 'I always remember the manifest. Each pipe was valued at £1,000 and that was 110 miles of pipe, so you can imagine the cost.'

The plan for the *Castora II* was to lay from Peterhead while an American barge from Brown and Root worked towards them. The barge was basically a flat deck 300-feet long, with a railing round it. Brian said, 'I have been there when the deck was about 30 feet from the sea and leaning over so much the water was almost coming over the side. When the weather was like that they dropped the pipe and picked it up again when it was calm.' He described the frenetic non-stop action. A section of pipe was fed below by what was called the firing line, down a large tunnel to six welding stations where different functions were performed. The pipe would be moved down a length by pulling the barge on its anchors. length. Then the next pipe moved on. The finished pipe entered the water down a protruding curved gantry, called a stinger after a wasp's sting. To continue the entomological analogy, the barge was like a giant spider laying one strand of its web as it crawled forward using its twelve, 15-ton stabilising anchors. The pipeline was lowered into a trench dredged or blasted out of the seabed. Before it was buried divers had to check that the welds were watertight.

Captain Carter was amazed at the speed of it all. 'One night I didn't

Plate 59.

The Italian pipe laying barge *Saipem-Castora II* reels out the pipes for the link with Forties field and the Cruden Bay landfall on the Aberdeenshire coast. *(BP plc)*

Plate 60.

Sometimes work on the rigs and platforms calls for the skills and nerve of a mountaineer, like painter Graeme Paterson and his mates who had to learn abseiling – or this man balancing above the waves on the Buchan field.

(BP plc)

think they would be doing anything because of the forecast, but they were laying so fast and the fuel boats were coming and going. Once you start offloading you have to take it all off, otherwise by the last tier it can be rolling a bit. We heard the Norwegian supply boat captain complaining because they had stopped offloading. The weather was picking up, so the captain wanted clear. "We have three joints of pipe left. Take it or else we have a problem." The voice came back, "We already have a so-and-so problem here." The whole shooting match had gone down, buckling the pipe. They lost about two months recovering it; divers had to cut the pipe, pull aside the damaged section and start all over again. They were always learning some very expensive lessons.'

The Herculean operation is one of the wonders of an industry constantly proving itself capable of conjuring up miracles. All the oil and gas fields are now linked to an array of terminals at Cruden Bay, the massive complex at St Fergus, at Nigg Bay, Flotta on Orkney and Sullom Voe on Shetland. Through the Far North Liquids and Associated Gas System (FLAGS), surplus gas is piped from fields near Brent, mostly to the Mossmorran gas separation plant in Fife. The alternative system is by tanker, either loading from a storage facility or directly from the seabed floating storage tankers (FPSOs), which can also produce and process the hydrocarbons. This system is used in difficult fields such as west of Shetland.

The contractors

By the time the bulk of the platforms were in place and producing in the 1980s, the majority of the personnel were non-operator staff. Occidental's Piper Alpha, for example, had twenty-eight separate contractor companies on board. The term contractor also applies to the worker and to the independent specialist consultants. The UK's biggest indigent oil service company, the Aberdeen-based Wood Group, is a leading contractor in an industry dominated by foreign companies, such as Brown and Root and their subsidiary Halliburton, Schlumberger and the various divisions of Baker Hughes. The local conglomerate grew with the industry from 1972, when young Ian Wood took over a modest family firm engaged in fishing, shipbuilding and marine engineering. Sir Ian, as he now is, has built a public group encompassing thirty companies in thirty countries, a staff of more than 9,000 and a turnover of over £600 million. More than half Wood's revenues now emanate from overseas. There are a number of other prominent Scots contractors, one is the Balmoral Group. Under the guidance of Jimmy Milne, the organisation, established in 1980 with five employees to design and manufacture glass-enforced plastic products, has grown to an international oil service company with a staff of 600, a manufacturing plant in Houston and an annual turnover of £100 million.

Out of the many thousands of experiences of contract oilmen, this is just one story. Like the Forth Railway Bridge, preserving the fabric of the miles of steelwork offshore is an interminable process for teams of painters. One veteran, Graeme Paterson from Aberdeen, works for the painting and scaffolding company Salamis. For the past twenty-eight years he has been a member of the group they call 'the Wild Bunch'. His first job was a Saturday crew change and he didn't get home at the weekends. The first opportunity he got to change to two and two, he grabbed it. 'My wages were probably 50 per cent higher than onshore, but looking back they weren't that great. But the money changed my life. Nowadays the onshore rates have caught up and it is the time off that is keeping guys offshore.'

The oil installation painters of the North Sea frequently have to perform like circus aerialists using a technique perfected in mountain climbing – abseiling – working from a web of ropes strung from the fretwork of steel structures, sometimes at heights of up to 400 feet above the surface of the water. Fortunately, heights have never bothered Graeme. 'There is no problem with the ropes; they are really safe. You have two ropes with all the attachments and a harness, tested to about 4 or 5 tons. What you have to watch is if you drop, throwing a shock load on a rope; a sudden thump.

THE OILMEN

That's what you worry about, but I've never been in a situation like that.' The painters were actually trained in abseiling through Chevron on a six-day course at Scota (now RGIT Montrose). 'That was horrendous, that was the worst bit. Because we had never used ropes before, we had no technique and my arms were in agony. We were trying to pull ourselves up using the upper body, the muscles of the arms. But rope work is all done with the big thigh muscles. You thought, "I'm never going to get through this." Then it got to the stage when you mastered the technique because your arms were knackered anyway. The certificate lasts for three years before you are reassessed.' Graeme described rigging the ropes. 'There's never anywhere you can't get to using ropes, compared to building several sets of scaffolding.' Across a broad stretch like a big bulkhead, the ropes are set up in tandem so the painter can move up and down, and all this several hundred feet above the decidedly unfriendly North Sea.

'I was on Mobil's Beryl Alpha in the 1980s. There was a diving boat alongside. Suddenly the boat shot underneath the exhaust on the platform and flipped the platform, knocking the entire superstructure off the boat. It was a weird sensation because we were up on scaffolding painting and the whole platform rocked. Your mind just couldn't take it in. The Beryl is a massive concrete three-legger. Then the platform flipped itself back.'

Graeme said Beryl was one of the better places to work in the early days. 'When you hit the BPs and Shells, they weren't the same. The Brent platforms' nickname was "the Killing Fields", not because of people dying because it was such hard work, and hard work with the Shell personnel as well. They were as bad as BP. But the Mobil guys had the same attitude as the contractors: "Here we go, out for a fortnight and then let's get home."'

Like many contract workers, the affable painter has been periodically faced with being paid off during cycles of boom and slump. 'Initially I worked on Conoco's Hutton tension leg platform for seven or eight years and then the Ninians for six years, so that was constant work and no problem. But two or three years back, everything just slowed down and dropped off and I was getting temporarily laid off. And they had you; you couldn't work and you couldn't find another job because they retained your P45. They only gave you a retainer to hold you on their books and you had to sign on the dole. I did that for probably six months and I had no idea if I was going to be working one week or not. That was no way to live. They could hold your P45 legally as a temporary lay-off, a system originally set up for fishermen. A lot of the contract companies cottoned on to this as a way of keeping their guys but not paying them. Nowadays if you are on temporary lay-off, you get an £85 payment which has to last thirteen weeks. If they can't give you work within that time they have to pay you off, make you redundant or find you work. But you can't go to them and say you have had enough and you are leaving. You would lose any service you had. I carry all my gear with me because I don't know if I am going to back to the same place – that is the life of the contractor. You are just bag and baggage. You go where you have to go.' Graeme admits that conditions have improved a great deal offshore. 'Nowadays, you have to do your training, every four years, but whereas it used to be three or five days, now it is one day. They say that's sufficient because you get the rest of your training offshore. That is crap. It's a different world these past ten, eleven, twelve years. Now it's all contractors that are actually running things, but the work is getting less and less every time.'

Ralph Stokes, a genial boilermaker from Australia, is now back home after working in platform construction yards and on installations offshore from the 1980s to 2000. His phlegmatic philosophy is that of a member of the North Sea supporting cast. 'Offshore was a lifestyle which suited me and I quite enjoyed it. I knew blokes who absolutely hated it. I was with one fella one time when we were getting on the chopper to go home. I said, "C'mon, cheer up you miserable so-and-so, we are going home." He said, "Oh yeah, but I have got to come back in a fortnight." There were a lot of people with that sort of attitude. Why work offshore if you feel like that? Get a job at home where you can be close to your family if that is what you want. Very few people are lucky enough to work in the ideal job or the ideal situation. You just have to make the best of what there is, I suppose.'

[7] First Oil Flows

The Hamilton Brothers' elderly semi-submersible *Transworld 58* may have been, in Michel Euillet's words, 'the biggest floating abortion in the North Sea', but it will forever occupy an important place in the history of the industry. On 11 June 1975 the rig, which had been ingeniously converted to a production facility linked to the wells on the seabed of the Argyll field, pumped to a tanker the first oil to be brought ashore in the UK.

For the small American company whose contract drilling rigs had been labouring for more than nine years in the northern sector it was a considerable coup. A week later, the tanker *Theogennitor* delivered 107,000 barrels of crude to a BP refinery on the Thames. The first of that company's huge Forties platforms was still four months adrift from coming onstream. Tony Benn, Labour's left-wing Secretary of State for Energy, who was regarded by the operators as either saint or sinner, turned the tap that transformed Britain into a global oil province. It was, he said, 'a day of national celebration and rejoicing'.

The Hamiltons had also achieved another less dramatic first: the utilisation of the *Transworld 58* as the first floating production system employed in the North Sea, a technique that took years to come into vogue. The system was later used on the giant semi-submersible, AH-100 for the Amerada Hess trio of fields, Hamish, Ivanhoe and Rob Roy, in the deep waters west of Shetland for Schiehallion and Foinaven, as well as by other multi-purpose 'floaters'. But the real colonisation of the North Sea began with a quartet of steel giants in the process of being planted in 1974, in BP's prize field, with its treasure trove of 1.8 billion barrels. Forties' forecasted peak flow of 400,000 barrels per day (later revised to 500,000) demanded facilities on a commensurate scale and with a direct route to the refineries by pipeline. The prodigious potential of this precious acreage of the sea, which Forties development manager Matt Linning described in 1972 as 'the jewel in BP's crown', was more than matched by the technological innovations and the massive investment package. BP set the audacious standard for the whole industry.

Plate 61.
The semi-submersible *Transworld 58* – described by a diver who worked from it as the 'biggest floating abortion in the North Sea'. Nevertheless, converted into a floating production system by Hamilton Brothers it became the first producer of oil from the northern sector. *(Niki)*

Since Block 21/10 had been allocated in November 1965 it had taken four years to find the oil and a further six years to reach production. Its development, which began in October 1971, demanded military precision in planning and execution. The daunting task is illustrated by the statistics. Four fixed steel production platforms, Forties Alpha, Bravo, Charlie and Delta, two built at Nigg in the Cromarty Firth and two on Teesside, had to be stationed in depths of 100 to 128 metres, 169 kilometres off Aberdeenshire. Each weighed approximately 27,500 tons, with three deck levels and a total height from seabed to the top of the drilling derrick of 210.3 metres. The platforms were floated out by rafts and then lifted into position in the heaviest-ever crane operations at sea. This was in 1974 and 1975. A fifth unmanned platform, Forties Echo, was added in 1986. A 32-inch diameter submarine pipeline laid to Cruden Bay, 30 miles north of Aberdeen, linked with a 36-inch diameter buried landline 209 kilometres to the BP refinery at Grangemouth on the Firth of Forth, where an oil stabilisation and gas processing plant, a tank farm, and a tanker-loading terminal had been built. The field was connected to the Aberdeen control centre by a

THE OILMEN

Plate 62.
Labour Energy Secretary, Tony Benn, turns a ceremonial tap on the oil tanker *Theogennitor*, moored in the Thames, and transforms the UK into the world's newest major oil province. *(Newsline Scotland)*

tropospheric scatter communications system. BP's initial estimate that it would take £370 million to construct their massive asset – which so shocked Grampian's convener, Maitland Mackie – was only for development; the whole capital outlay would be more than £1 billion. That initial £370 million was borrowed from sixty-six British and foreign banks, the biggest bank loan then authorised in Britain. BP Development general manager, Basil Butler, said in 1978 that Forties gave the lie to the idea that Britain could no longer handle major international projects. 'We acknowledge the technical assistance we have had from other companies in developing the field, but this was primarily a British achievement.'

In terms of human endeavour it was also a colossal achievement, remembered by those who took part in it as exhilarating. Ted Roberts, a bluff Yorkshireman who earned the nickname 'The Godfather', was operations OIM on Forties Alpha and the first to set foot on the completed production platform. 'It was so exciting. You didn't even feel tired. I mean you did, but you didn't realise it because you were so involved. My fingers used to tingle. You had a shower, jumped into bed and three or four hours later you wanted to get up and go after it again.'

The offshore adventure began when the carcass of the platform, initially known as Graythorpe One, hove into view on 4 July 1974 towards a small armada of boats and barges. Master of the command vessel, the *BP Oil Producer*, John Carter from Caithness, had a ringside seat as history was made and he was also there when Highland One (Forties Charlie) was

THE OILMEN

Plate 63.
Forties project manager Matt Linning issues instructions from the control vessel, *BP Producer*, as the installation of the first platform reaches the crucial crashdown stage. Watching the operation, BP Chairman, Sir Eric Drake. *(BP plc)*

installed a year later. There had been six weeks of careful preparation and rehearsal for the great day. 'The *Producer* was basically a supply boat fitted with a self-contained control unit to activate the signals for ballasting and installation. The engineers were from Brown and Root, Americans, highly intelligent, smashing guys, some of them from NASA. This was a first, very high-tech for the time, using a pulse rate and telemetry.' The concept of a count down came directly from the space programme.

Britain's first offshore platform was towed out from Middlesbrough by two tugs, one German and one Dutch. 'At that time there was friendly rivalry between the two crews; both their teams were in the World Cup final that week. We went on station with Matt Lining from BP and Roy Jenkins, the

Brown and Root boss. The BP chairman, Sir Eric Drake, arrived by helicopter on to one of the derrick barges. It was a bonny day and everything went fine.' Matt Linning, who became general manager of BP Petroleum Development, described the field 'as the sweetest thing you could get. An oilman's dream, the only complete system in the North Sea bringing oil from wellhead to its own refinery'. Increasing the tension was the fact that Graythorpe was installed in the full glare of the international media. 'When the platform was emplaced successfully the very first telex message of congratulation was from the bankers. After all, they were the men lending us all that money.'

Captain Carter described the operation. 'The platform was slowly tipped up in stages from its pontoon. The crucial bit was from about 20 degrees to about 60 degrees – the crashdown – but there were flotation buoys to prevent her wobbling and huge hydraulic clamps were released when she was piled in.' Suddenly, the controllers became aware of an unwanted spectator. 'A Russian trawler with more aerials than a porcupine has quills. The Navy chased it away. If it had come between the command vessel and the jacket it would have interfered with the signals, but everything went smoothly. Then the two derrick barges, *Thor*, owned by Hereema, and *Hercules*, from Brown and Root, moved in with massive steam hammers, 6 feet in diameter and 120 feet high, and dominating the barge, the great crane. They had to put in four or six piles to each leg, before they could release the flotation. What upset us was the tremendous noise of both barges knocking in the piles.' The risk-laden exercise was a perfect textbook operation. Linning, the organisational and engineering genius, talked through the remarkable feat in an interview to commemorate Forties' billionth barrel of oil. 'There were really hairy moments with the weather and heavy seas running and the fact no one had ever done what we were trying to do. We took risks – most of them calculated. But sometimes nobody really knew just what was involved at the time. We just went ahead and did what was needed.'

Despite the conditions, the whole massive Forties operation was carried out over five years without any major accident or fatality. Yet Linning himself could have been one of the statistics. He admitted later he had narrowly escaped drowning during construction. In total darkness, he was making his way from a derrick barge to the platform when he fell into the sea. Fortunately he was seen and rescued. He had already survived two earlier air mishaps in the Persian Gulf. Tragically, the architect of Forties, knighted for his achievements, died after a fall during retirement.

The next step involved construction and commissioning as the crews battled to bring the giant field on stream ready for the first oil to flow down

Plate 64.
Master of the control vessel *BP Producer*, Captain John Carter maintains her position (Plate 65) as Graythorpe One topples from her floation tanks.

Plate 65

the pipeline, which had been crawling towards its underwater rendezvous halfway between Forties and Cruden Bay. For Jim Soutar, from the Merchant Navy, his first job as a mechanical supervisor was in a portacabin onshore. 'Recruitment for Forties started in earnest in August 1974. The intention was to be offshore by Christmas. I joined in September; Christmas came and went and it wasn't until around March or April 1975 we started going offshore in any numbers, and the first oil from Forties was that September. We spent our time in a so-called 'think tank', poring over manuals and drawings. Once

Plate 66.
The painstaking process of lifting the modules on to the jacket underway on Forties Field. *(BP plc)*

Plate 67.
Hereema's mighty crane barge *Thor*, one of the two giant tugs involved in the construction of Forties Alpha. They piled in the platform with huge hammers. *(BP plc)*

offshore we worked with the people doing the commissioning and picked it up as we went along. The operations managers were like myself – they had been in the tankers for Shell as well as BP, it was pretty much the blind leading the blind – amazing. There were very few BP core people. The platform was like a ship, so someone with command experience was the best choice then. I went from mechanical supervisor in a year to operations manager, the second in command, and then OIM.'

It was estimated it took 20,000 people to bring Forties to production. Ted Roberts was responsible for recruiting. 'The men I worked with during those years were the finest team BP ever put together. We had something special.' They represented an incredible range of occupations and training. On one platform alone there were Merchant Navy seaman, a crane driver, a cocktail barman, a policeman, a tanker captain, a pharmaceutical salesman,

THE OILMEN

Plate 68.
And the finished product – Forties Alpha on stream. *(BP plc)*

an RAF electrician, an RAF nurse, a garage owner, a trawl skipper, a trawl fitter, a lorry driver, a mill maintenance engineer, a paratrooper, a cold storeman, a restaurant owner, a carpenter, a builder, a fireman, a estate factor and a gold miner. They came from central Scotland, north and north-east England and the Midlands, but, in the spirit of the industry, there was also a catholic mix of nationalities.

One senior deck official, Eric Strachan, recalls that a large number of Spanish and Mexican workers were brought in for the construction phase: 'There were so many of them living in temporary accommodation we named the cabins "Inca City". They were really hard workers and put in long hours to save up the money to send home.' According to Jim Souter, the Americans used these workers, who were employed by UMEC, Brown and Root's manual labour division, because they were cheaper. 'These guys used to work

Plate 69.
Not even the frenzied activity of BP's Forties pipeline construction a few hundred yards away on the beach puts the golfer of this putt on the Cruden Bay course. *(BP plc)*

three months offshore without a break. The hours they worked would not be tolerated now. It was amazing there were no serious accidents. Pipe fitters would work 18 hours a day on top of production modules and they would fall off – probably asleep.' Jim said that it was only towards the end when Forties Delta was commissioned that the workforce underwent a change. 'The senior engineer for Brown and Root didn't like Americans. So he replaced them with British people.'

Former fellow OIM Andy Lawrie maintained there was real pressure to produce oil in 1974. A Scot, he had served as an engineer in BP tankers and

[141]

Plate 70.
Nightfall – and progress continues on the Forties pipline in a 24-hour-seven-days-a-week race from land to meet the line snaking out from the field. *(BP plc)*

as a technical superintendent in the Middle East. 'We were producing oil and gas with a not very well trained workforce while we were still constructing the platform, so there was heavy engineering, welding, burning, going on – the kind of situation you didn't really want.' The safety boats were small trawlers. 'That filled the letter of the law but certainly not its spirit. Safety in such a volatile environment was a constant concern for the OIM. There were a number of fires on the platform and I got a call from my boss. "Heads are going to have to roll out there and I am coming out to speak to you." I was certainly not going to be bullied, but there were areas nobody would take responsibility for, so there were potential fire hazards. During the third fire I had stopped production, so the pressure was on and I felt completely alone. There was a very cold response from the BP representative: "I hope the board will be understanding." I ignored it, but those were the kind of tremendous pressures. I would say we built a Frankenstein out there and management weren't aware of it. When they realised the work was taking longer than anticipated, they built a massive new accommodation block with two-man cabins. The point was to keep the oil going.'

Fellow OIM Jim Souter said when first oil came from Forties Alpha

Plate 71.
The crucial moment – the Queen inaugurates the Forties Field and starts the first oil flowing down through Scotland to the refinery in Grangemouth. *(BP plc)*

through Charlie, it was a great achievement. 'The senior guy was running round with about six radios hanging from him shouting orders here, there and everywhere. So there was tension, particularly when some startups didn't go as planned. After the oil started flowing there was always a challenge. After a few years some folk became disenchanted, caught on a financial merry-go-round and couldn't get off. To most people, however, it was probably among the best times in their working lives.'

Finally, the field was inaugurated in September 1975, by Her Majesty the Queen in a spectacular ceremony at Dyce in Aberdeen, followed by a lavish party, said to cost BP £30,000 – a considerable sum even by high-spending oil entertainment standards. By then, a long ragged line of platforms from the far north to the central and mid-North Sea, was beginning to transform the empty windswept wastes. Five more licensing rounds up to 1985 unleashed a further 337 blocks – 157 companies were successful in the Eighth Round in 1980. The discoveries began to mount up while Forties was being created; there were thirty-seven significant finds, twenty-five in 1975 alone – a figure never again achieved. More than 600 wells were drilled over the following decade, averaging fifteen a year towards a total of 156

THE OILMEN

Plate 72.
The Clair field, under construction – BP's latest asset in the promising sector on the Atlantic Margin, West of Shetland. Discovered in 1977, it is only now being developed more than quarter of a century on. *(BP plc)*

commercial wells. Apart from Statfiord, discovered in 1974, and whose reserves of 3.5 billion barrels proved it to be the biggest field in the North Sea, it appeared that the 'elephants' such as Forties and Brent had all been located.

Nevertheless, a number of fair-sized fields with now – familiar names were found. Among them were Beryl, South and North Cormorant, Dunlin, Hutton, Heather, Magnus, Brae and Fulmar. Publishing conglomerate, the Thomson Organisation, had a stake in Occidental's Piper and Claymore fields and were represented by Alistair Dunnett, former editor of their newspaper, *The Scotsman*. The foray by Britoil had also begun to pay dividends. Their first find was Amethyst, in the southern gas fields, followed in the north by Deveron and then the big field, Thistle. The decade's discoveries ranged across the whole of the UKCS, with BP's Magnus the furthest north, 124 miles out in the east Shetland basin. But drillers also ventured into the deeper waters of the Atlantic, on the opposite side of mainland Shetland. There, in 1977, BP drilled into a field they named Clair. The technical problems of extraction posed by that find (an estimated 4 billion barrels of heavy crude), meant that it had to lie untapped for more than twenty years, until the necessary technology was developed. Clair also opened up the prospect of a new and exciting oil patch to inspire a revival into the fourth decade of a mature UK oil industry.

In Armand Hammer's golden blocks in the Moray Firth, off the shoulder of Scotland, Piper and Claymore were discovered a year apart. Two years later, in 1976, there was a discovery 12 miles from the Easter Ross coast, the nearest field to mainland Scotland. International oil consultant Alex Barnard

Plate 73.
Easter Ross's own field – Beatrice – only twelve miles off the coast. The locals wanted to name her Brora, but the American owner, the legendary T Boone Pickens, had other ideas. *(BP plc)*

was working for P&O, who had a stake in the block along with Mesa, a company owned by the legendary American oilman, T. Boone Pickens. 'There were two companies who had a crack at block 11/30, but the wells were dry. I asked an amazing geologist I knew, Professor Cliff Potter, to look at the data. He explained, "There are Devonian and Callovian sediments which are similar, but the Callovian is dry. I think they stopped at the Callovian." It was worth a try and Potter was absolutely right – we found the field 500 feet deeper, in the Devonian field. Not very good oil, but prolific. The field was sold to Britoil and became BP's when the companies merged.' It is now operated by the largest independent, Canadian company, Talisman.

The people of Easter Ross had adopted the field as their own and for weeks there was speculation about its name. Former industrial editor at *The Press and Journal*, Dick Mutch, recalls that the locals had decided on Brora –

THE OILMEN

Plate 74.
Prime Minister Margaret Thatcher unveils a plaque in Britannic House, London, to inaugurate BP's Magnus Field, north east of Shetland. Her Government were to ease the taxation shackles on the oil companies.
(BP plc)

after the nearest landfall. 'That year, 1976, I met Boone Pickens at the Houston oil conference and I told him about the great local interest in what he was going to call the oilfield. He was puzzled. "I have named her already – 'Beaaatrice'." That was how he pronounced it. "That is mah wife's name." I sent the story and all sorts of rows broke out over this American upstart who dared to name their oilfield.' But Beatrice it has remained. The naming of oilfields was, in fact a deliberate exercise in the kind of public relations the shrewd international companies practised to sanitise an industry which even then did not always enjoy public approval.

THE OILMEN

Following the success of Forties, exploration suffered two relatively lean years while production plans were put on hold by companies alarmed by the fiscal situation of the late 1970s. When the era of Thatcherism dawned in 1979 the petroleum revenue tax was raised from 45 to 60 per cent. At one point in 1982, UKOOA claimed thirty fields were at risk for tax reasons and lobbied hard for change, as they felt investment was being inhibited. Surprisingly, thirty years later a former oil company boss conceded that on the whole the Government had tailored the tax structure to suit demands in the early stages. Bill Schmoe, vice-president of Conoco UK, was president of UKOOA during that critical period. 'I think the UK Government did an absolutely superb job to encourage exploration and development. Maybe they gave up a little income at the early stages, compared to some other countries, but they really benefited in the long run.' Another former Conoco boss and president of UKOOA, Harry Sager, had another view. 'PRT didn't leave anything after the projects were paid out, particularly if the price of oil was going down. A good bit of the trouble I had with the Government was trying to make it flexible enough for us to look ahead. They finally did fix that.' And that was in the 1983 Budget. According to a UKOOA document in 1986, 'Exploration has jumped to record levels and development consents have reached double figures since these changes.' As the production curve continued upwards from 10 million tonnes in 1970 to 180 million tonnes in 1985, revenue from taxes and royalties also increased, from £2.3 million in 1979–80 to a peak of £12.2 million in 1984–5, before tumbling drastically to £4.7 million in the 1986–87 crisis.

By 1985, thirty-one more fields were in production with seven more under development. For the first time (1977), more than a million barrels of oil a day were produced, and in 1981 production exceeded consumption; Britain had become self-sufficient in oil. While production roared ahead, huge sums continued to be poured back into the North Sea. In the 1960s, the cost of developing the new gas fields was estimated at more than £250 million – a figure which was far surpassed in actuality. But if the northern sector was breathtaking, from 1965 to 1985 exploration cost over £11 billion in a total expenditure of £28 billion. Those intense and busy years from onstream Forties to the big oil slump in the mid-1980s were the most exhilarating in the history of the UK industry. Martin Reekie was part of the wave of recruits rushed in to meet the dramatic expansion 'It was very much boom when I started with Shell in 1980; huge fields being found, discoveries all the time, developments, money ploughed into it, money no object, and you couldn't really see it ending.' Martin worked on the Brent Bravo and Delta installations. Bravo had been up and running for four years, the first of

Plate 75.
Bill Schmoe, former boss of Conoco and president of UKOOA in the 1970s, thought the UK Government did a superb job with the taxation regime. *(Conoco Phillips)*

[147]

THE OILMEN

the four platforms – three concrete and one steel – 900 feet high, standing in 460 feet of water. The investment on the Brent System was equally awesome: expenditure (in 1996 money) £11 billion; operating costs £8 billion; tax and royalties £13.7 billion. Bravo produced initially to a storage spar buoy facility, which was 463 feet high and could handle a quarter of a million barrels a day directly from the field, feeding tankers which transported the oil to the shore. Ultimately, oil from Brent, from Conoco's Murchison and Hutton, and from Britoil's Thistle, was pumped to Sullom Voe in Shetland via the Brent pipeline system. Sullum Voe, which received its first oil from Shell's Dunlin field in 1978, means 'a place in the sun' in Norse. And the sun most definitely shone for a 25-year spell on the shrewd Shetlanders, who had driven a hard bargain for a generous long-term share of the revenues from owners of the terminal, which was operated by BP. The gas from Brent BP's Magnus, Unocal's Heather, Chevron's Ninian and Statoil's Statfiord eventually followed the much-delayed FLAGs pipeline to the St Fergus gas separation and processing plant, opened in 1982. Seventeen oil companies had agreed to build the Brent line as a common carrier.

A young production operations representative with Shell, John Wils, worked on the early innovative Brent gas compression system, installed before St Fergus and the pipeline were ready. 'The field had a very high gas/oil ratio and there were government restrictions on flaring. Brent faced considerable losses unless we got some gas compression facilities quickly. The job was undertaken by Esso, the partners of Shell. So I worked in Houston as operations rep in the Dresser factory, where they were building the world's biggest offshore reciprocating compressors. These modules were actually floated straight out into the North Sea. You can imagine the engineering effort. It was incredible – the interface of electrical instrumentation and pipework and getting the compressors onstream to inject the gas. I led a production team in the start up and in commissioning these facilities. It was a world first. Eventually the gas operation was transferred into the pipeline system, which was always intended.'

Ex-patriate Charlie Brown was transferred by Shell in 1975 from Nigeria, where he had been barge engineer on a jack-up rig. He was involved in the Auk Alpha jacket installation which ran into problems. 'The loading facility, an exposed location single buoy, had sunk, but we recovered it and got it floating again. They piled up the jacket and the modules, built in Holland, came out in barges. These were installed and Auk was up and running. It was all so different to what I had been accustomed to overseas.' After a spell as head of maintenance at Shell's base in Aberdeen, he went on the new OIM training programme before going offshore. Then, in his own words, 'along

Plate 76.
Brent Delta – one of four platforms on Shell's 'elephant' field, west of Shetland. The field became the central point of a spider's web of pipelines that gathered gas and oil from other fields and fed them to terminals on Sullom Voe and St Fergus. *(Shell International Ltd)*

Plate 77.
The biggest moveable man-made structure – the gigantic jacket that was to become Ninian Central, under construction at its birthplace in the fabrication yard at Kishorn in Wester Ross. (*CNR International*)

came Dunlin, to my mind the jewel in Shell's crown'. This oil and gas field, discovered in 1973, with estimated recoverable reserves of 50 million tonnes of oil, was to be part of the Brent system. The concrete base was built in Rotterdam, towed to Norway and finished off there before being hauled out to the field. Charlie was among the Shell employees who celebrated Dunlin's quarter century of production. 'I was the first OIM on board and I went right through to first oil – drilling and construction all at the same time. As a platform it was lovely. We had learned a lot from the Brent platforms in design, layout, technology and construction methods; things were more finished in the yards. We had some bad experiences on the early platforms, but that was part of the learning curve. We came through it pretty well.' A broad mix of nationalities brought the field to production in 1978. 'We had Spaniards, German drillers, a French company doing the anti-liquefaction plant and, of course, a few Americans. The hook-up contract was with

THE OILMEN

McDermott, and the project manager was a Welsh Canadian. But they all blended very well, which was part of Dunlin's success – drilling, operations, maintenance, construction, and project manager, all blending together. It was exciting. Life was good on board.'

To the west of the Brent field was Chevron's huge Ninian reservoir, the third largest in the UKCS. Like almost every other operation in the North Sea, bringing Ninian's three big platforms, Central, Southern and Northern, to production was a daunting logistical operation. The Southern and Northern platforms were made of steel but the Central installation was concrete – the largest moveable object then known – built at a cost of £60 million at Loch Kishorn in Wester Ross.

Three former OIMs retain vivid memories of those hectic days. Alan Higgins from Rugby in Derbyshire was on all three platforms throughout the hook-up and to first oil. His background was in oil tankers, first for BP and then Chevron, ending in the Middle East as a marine superintendent. In 1978 he trained as an OIM and went to Ninian Central during the organised chaos of the hook-up. 'It was like an anthill, with thousands of people, and because it stood on a circular concrete tower, it was very difficult to pull a semi-submersible alongside a gangway. So we helicoptered the workforce from the accommodation vessel, the *Boglin Dolphin*, a couple of miles away. Twenty or thirty helicopters shuttling twenty-five crew; it was incredible. Money was no object. On foggy or stormy nights when the helicopters couldn't come, what the hell could you do with all these people? You gave them a sleeping bag and they had to find their own space.' Alan found the sheer size of the platform hard to grasp. 'Central was an original concept, built as one big central column with tanks underneath, and there was a 13-metre-deep skirt like a steel pastry cutter that was fixed into the seabed. The ship I had come off was nearly 300,000 tons – here was this thing, twice the size, and I thought, "My God, that is incredible."'

Another OIM, local man Alex Riddell, also arrived by the Merchant Navy route, working first in supply boats before being hired by Chevron as an offshore operations supervisor on Ninian South during construction. 'Things didn't go to plan and as it started to get near the first oil day, people began to get frantic because the creditors were saying, "Wait a minute, we have £20 million in this project. It's come the time I want to get the money back." Unfortunately, there are two worlds out there. It's like a pendulum. At the start, the hook-up contractor is the king, then midway through it's the operator's call and he says, "Come on, get this finished, we have to produce first oil." I was operations supervisor and after first oil, maintenance supervisor.'

It was to Ninian Southern that Alan Higgins was permanently assigned for three years, arriving four months before production. 'I was there when first oil came. Like all these things, it all happened in the middle of the night. We opened the first well and the separators filled up. Literally everybody was hanging over the rails waiting for the flare, and we were the first in the field to have our flare lit.' Again, the sheer volume of manpower impressed the OIM. 'There were probably 800 people streaming across from the accommodation semi – like Port Glasgow years ago; tenements on one side of the street and shipyards on the other, and at 6.30 a.m., the traffic had to stop and all these people came out of the tenements and straight in the gates. Then Central came on line, and eventually Northern. Just the nerve to think they could do it. Northern more or less fell off the back of a barge and sank within a metre of where it was supposed to be. Some guy sat down and calculated this – it was magic.'

On Ninian Northern at that time was another OIM, John Nielsen, a former Royal Navy engineering officer who had answered a Chevron newspaper advertisement for a job in the Ninian Field. Readying the Northern platform for production took eighteen months. 'It was hectic, all the usual hook-up problems and you had Lloyds checking to see if it all met requirements. They just threw money at it, as the quicker it was built the sooner they would get their money back. I never knew from one shift to the next where I would be sleeping. We just shared. The accommodation vessel – the heavy lift barge – was only there during construction. The first modules out were the accommodation ones. OIMs weren't any different from anybody else. If we were lucky we had a cabin – but not during hook-up. There were people who came out and didn't ever actually do anything – just went from one place to another. There were so many.'

Alan Higgins ultimately spent three years, seven days on and seven off, on Ninian Northern. 'We still suffered a lot from our lords and masters in Aberdeen who were largely American, because it was an American company. Occasionally, something would happen and the boss would come on the phone. You would tell him it wasn't the best or safest way of doing things and you would get, "Goddam, I'm the boss," and you were expected to toe the line just as the American supervisors would do. But we were put out there as OIMs to uphold the law and the Health and Safety regulations, all that kind of thing. It was just an extension of what I'd been doing before. All my six colleagues on the Ninian platforms were the same. They were all Royal or Merchant Navy.' In some companies, a strong production manager onshore may demand that all big decisions come through him, according to Newcastle man, Roger Ramshaw. A petroleum engineer with Conoco, he

THE OILMEN

retired as managing director of ConocoPhillips UK. He disagreed with that policy. 'At platform level the guys must feel empowered to just shut it down. You go to something like the Cullen Inquiry and hear about the indecision, and when you see the re-enactment, it makes you squirm and shout, "No, just shut it down!"'

As the industry developed across the basins it had constantly to reinvent itself to meet the differing challenges encountered with each new field. Roger Ramshaw worked on drilling rigs in the East Shetland basin at that time. 'Generally, it is the technology that makes the difference. Organisational stuff is superficial compared to the effect of, say, subsea well completions. That is second nature now. Then, you did something, not on the basis of "Oh, we are going to have a research project," it was "I have to do this by tomorrow." You would set something out on paper and get someone in the machine shop down the road and it was done.'

Young driller Roy Wilson found himself totally immersed in the technological changes when he left ARCO to work for the growing American oil operator, Mobil, in 1975. He wanted to go where there were new ideas. 'Mobil had already installed Beryl Alpha, the first Condeep concrete platform ever built. History now can testify the platform was put in the wrong place, not deliberately. When they drilled the first wells, which they thought were the meat of the field, they found out that reservoir was no good. It took a year or two before they found the good wells.' He was assigned to the installation of the very first electro-hydraulic remote-controlled subsea well in the North Sea. 'Developed by Seal, which was set up by some of the big companies, it was an all-singing, all-dancing electronic scientists' dream, but a practical man's nightmare. Controlled by telemetry, it was originally designed to be in a capsule in seriously deep water. In 1975, people had all kinds of major headaches, trying to get it in place and working, but I thought it was wonderful and couldn't understand why no one else shared my enthusiasm. It saved the Beryl platform in the first year or two. In addition, the subsea well was producing oil to power the turbines. It was an unhappy time until we hit the payload so to speak – that period cost a lot of money.'

Other companies were also working on their versions of subsea completions, which were satellite production and control packages on the seabed tied back to the mother platform. Shell were first with an attempt on Brent which wasn't totally successful. Martin Reekie was on the *Stadrill*, when they tried again in advance of far-sighted plans to install an underwater manifold through which wells could be drilled and tied back to Cormorant Alpha. 'It was their first successful tieback.' The project, which cost £360 million, was the key to the next generation of developments. 'It was so

Plate 78.
The Murchison platform in the 1970s when the field was developed by the American company Conoco. The asset is now owned by Kerr McGee. *(James Fitzpatrick)*

important,' claimed Martin, 'well over half Shell Expro's production comes from subsea now.'

Platform design had also to be at the forefront of technological change. Conoco's Hutton platform is a prime example. The field, discovered in 1973, was named after eighteenth-century scientist James Hutton, widely regarded as the father of geology. The American oilmen had just brought another field, Murchison, to production successfully, but Hutton was different – a milestone in the history of offshore technology with its unconventional and highly original design. It was the world's first tension leg platform (TLP), the hull built at Nigg and the deck at Ardersier. The design actually came before the decision to develop the field, says Harry Sager. 'Hutton was a bit troublesome geologically. The pay sand thickness was not very uniform and it was faulted up quite a bit. So it had been put on the back burner several different times. We had a very strong engineering group in Houston who had this concept pretty well thought out to use in deeper water, and they figured somewhere in the northern North Sea would be ideal.' As the first TLP, Hutton meant a far greater engineering involvement by Conoco. Tom Marr, the project engineer, said, 'I know why they did it. If we could show it worked, then we were damned well fixed to go into the deep water. By deep water we are talking about the 1,000-foot range – 750 feet is the limit for a fixed structure. So we were looking further in the North Sea, all the places you have deep water. So Hutton was gearing up and that success was to be taken into greater depths.'

Jack Marshall, Conoco's vice president of production and exploration said Hutton was a tough call. 'We finally got our partners and our own company to agree it would make money. It wouldn't be an experimental loss, and in the process we were going to learn an awful lot. It was a very bold decision.' Harry Sager said, 'If we had been looking at it in 1985, we would probably not have done it with any platform.' But the doubts of the six partners in the enterprise, Gulf Oil, Amoco, Enterprise, Mobil, Amerada Hess, and Oryx – the company who later bought the field – were overcome. It was a curiously shaped 23,000-ton structure, with six steel columns and sixteen tension legs which acted as tethers from the hull to a template on the seabed. The buoyancy of the platform put the legs under tension. The two sections were connected in the Firth and then hauled out by tugs on its eight-week journey to the East Shetland basin. Three weeks after hook-up it began producing oil; a conventional platform would have taken at least a year. Conoco had meanwhile become part of Du Pont, who were acting as a white knight to stave off a takeover by Mobil.

Since his retirement as director and general manager for Conoco's UK

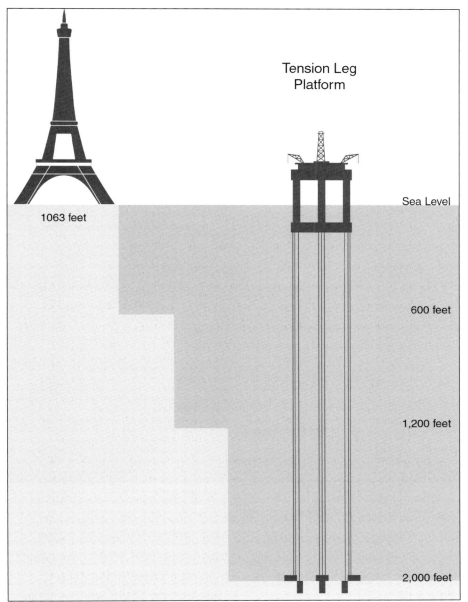

Figure 4.
The revolutionary design for a tension leg platform built for Conoco's Hutton field. The style is now commonly used in deep water in the Mexican Gulf. *(Greybardesign)*

projects, Tom Marr has travelled the world preaching the merits of the TLP design. Conoco, in fact, produced another one for the Norwegian Heidrun field, but with a concrete hull instead of steel. Jack Marshall said that now all the platforms offshore in the Gulf in 4,000 and 5,000 feet of water are derivatives of what Conoco produced more than twenty years ago. The Hutton TLP has since been successfully decommissioned by its owners, Oryx (part of Kerr Magee, who also now operate Murchison and Ninian) and was sold off in 2002 to a company operating similar platforms in the Gulf of Mexico.

Plate 79.
Sunset over the unique Hutton TLP – now dismantled and sold to an operator in the Gulf of Mexico.
(James Fitzpatrick)

Back in the 1980s, there were few signs of a looming crisis. A number of new discoveries appeared – mostly small and medium-sized, while some notable fields reached first oil. But there were also signs companies were beginning to scrutinise the prohibitive costs and the new fields – an example was Shell's Fulmar – were coming on stream more cheaply. Managing director Dr John Jennings forecast, 'The future of the North Sea development will be largely with smaller fields, costing very much more than their predecessors for every barrel gained. To be economically viable, they will need tight control of cost and innovative technological systems such as Fulmar has achieved.'

Then came 1986. Like the rest of the world's oil-producing countries, the North Sea had suffered in 1972 when the Arab–Israeli conflict forced the global price up. However, it was the panic of 1986, when the benchmark price crashed from $30 to $10 a barrel, which defined the future pattern for the industry. The events of that traumatic year demonstrate vividly the dependency of the Treasury on oil and gas revenues; they also provoked a financial reaction in the industry which many people believe had disastrous repercussions.

In 1985 – the zenith for the industry – levels of exploration and production were the highest in UKCS history. The Minister of State for Energy, Alick Buchanan Smith, proclaimed, 'The rate at which oil and gas is being discovered is unprecedented anywhere else in the world.' The success rate was also greater than in any previous year. At a conference in Leith, petroleum consultant A. Gaynor said, 'The North Sea is on the brink of a boom that could be sustained through the next decade and into the twenty-first century.' Revenue, valued at some £2.5 million per hour, had become integral to the economies of the United Kingdom, Scotland and the Grampian Region – far too dependent, in the belief of a senior Scottish politician, who in the thick of the first abortive devolution battle in 1972 had coined the Scottish National Party's most memorable but abortive political slogan: 'It's Scotland's Oil'. Dundee solicitor Gordon Wilson, leader of the party in the 1980s, wrote in *The Scotsman,* 'The plain truth is that the UK economy is hooked on oil and that there is a complex and fragile relationship between oil production, prices, the balance of trade, oil revenues and the value of sterling which is treated internationally as a petro-currency. A slip back in any of these elements could promote a major crisis.'

In January 1986 those very circumstances conspired to force a dramatic price plunge to $10. Trade figures showed the cost to the economy; £1.6 billion on the balance of payments, 'equivalent to the value of Britain's annual exports to the rest of the European Community'. The problem, inevitably, lay with the ten volatile states who formed the global energy cartel, the Organisation of Petroleum Exporting Countries (OPEC), and who held 66 per cent of the world's oil reserves. Neither Britain nor Norway were members. OPEC was in contentious disarray. Adherence to their quota system controlling world oil production was virtually non-existent; Iran and Iraq, in particular, had increased their output to finance their arms race. Finally, the biggest producer, the Saudis, abandoned the artificially high oil price and increased production. The price fell dramatically.

The industry and the Government nevertheless masked their concerns with breezy optimism. On the day Brent crude dropped below $18 a barrel, Iran's main oil export terminal was flattened by Iraqi warplanes and OPEC leaders met in Geneva to discuss the crisis, Energy Secretary Peter Walker made a major policy speech about the North Sea. According to a financial journalist, 'Not once did he refer to the fact that the price of Britain's crude has fallen by 40 per cent in the past two months, nor to the growing fear that unless prices

Figure 5.
The various designs of fixed and mobile installations employed in the production of oil and gas since the 1970s. Apart from the TLP (seen on a previous page), there are six different types: the flotating production system; the gravity platform; the guyed tower, the steel piled platform; the tanker; the steel piled installation with the now ubiquitous subsea completion system – the underwater manifold which can handle tie-backs from numbers of wells. The choice of design depends on the size of field, depth of water and ultimately the economics of scale. (*Greybardesign*)

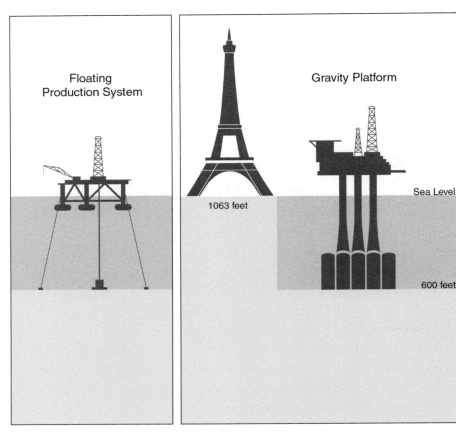

stage a spectacular recovery, there will be little or no development of the North Sea beyond what has already been set in train.' Mr Walker told the Royal Society in London, 'In energy we can boast that much has been achieved, but what has been achieved is nothing compared with the opportunities that lie ahead.' Oil industry journalist Axel Busch calculated, however, that when the price of oil dropped by $10, the British Exchequer lost £500 million in revenue. In reality, the Government were desperately trying to stem panic and reassure investors. Alick Buchanan Smith warned oilmen at the annual Offshore Technology Conference in Texas not to overreact: 'A heavy price has been paid for overreacting at the top of the oil cycle, let us not compound that mistake by over-reacting now as it reaches the bottom.' Shell UK's managing director of exploration and production, Peter Everett, sounded a similar note when he told concerned oil construction yard executives at a seminar in Strathclyde University, 'It seems to be the season for politicians to remind the oil industry that they are in a long-term business and should not be deflected in their planning by present uncertainties. I recommend that investing time and money . . . must continue at a healthy pace.'

However, the crisis was being spelled out in stark statistics; proceeds

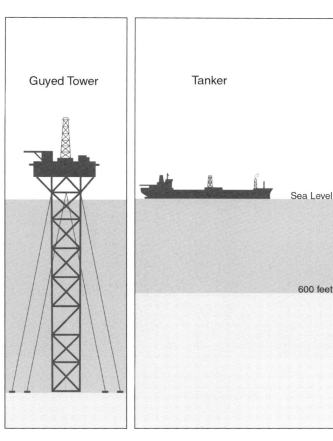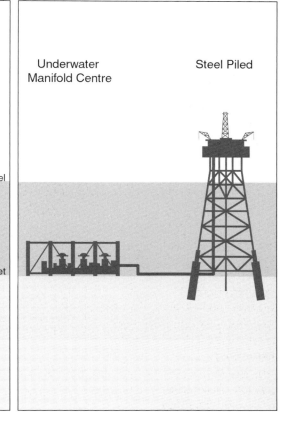

from sales of North Sea oil fell by more than half – from £19.7 billion in 1985 to £9.3 billion (gas sales on the other hand increased from £1.7 billion to £1.9 billion). The oil companies, principally the majors, began to pull back; BP's profits in the first quarter of 1986 dropped by £718 million and the company cut exploration expenditure by 20 per cent; Shell's profits were down by £300 million and spending on exploration fell from £5,700 million to £5,000 million; the number of significant new finds dropped from twenty-five to fourteen; in Grampian 10,000 jobs were shed; unemployment, which had stood at 3 per cent (5,300 people), rose to 11.2 per cent (25,000) by January 1987; spending on support industries dropped from £15.498 million (1981–1985) to £10.476 million; the industry's flow of funds had reduced by some £3 billion between 1985–6; service companies were hit hard – forty fewer supply vessels were at sea; the workload in the oil construction yards went down by 60 per cent; drilling companies' commissions dried up and rental rates for drilling rigs fell from $100,000 to $10,000, leaving eighteen redundant structures stacked up off the Firth of Forth, the Moray Firth and Aberdeen. The North Sea industry was in deep trouble.

Pressure mounted on the Government from politicians, the City and

Plate 80.

Offshore – the workforce and the technology that produce Britain's oil and gas wealth.

(Greybardesign)

OPEC to cut North Sea production. Norway said they would, if Britain would, but Chancellor Lawson dismissed appeals for tax changes and forcibly expressed his Government's clear policy in his 1986 Budget speech: 'There is no question whatsoever and never has been any question of the UK cutting back its oil production to secure a higher oil price. The whole outstanding success of the North Sea has been based on the fact that it is the freest oil province in the world, in which decisions on levels of output are a matter for the companies and not the Government.'

This was no consolation to those trapped at the epicenter. In Grampian, where nearly 80 per cent of the Scottish oil industry was deeply embedded, there was a domino effect. First to go were many of the 13,000 migrant workers from across the Border, taking their spending power with them; then the majority of 3,000 American incomers, most in the higher earning brackets. They went leaving keys, houses and mortgages behind. A phenomenally large number of houses were marooned, with nearly 4,000 homes for sale, increasing at a rate of 100 a week. The recession was rebounding on the onshore businesses which had become too dependent on the trade. The only sector prospering was Aberdeen Harbour, where the 2.7 million total supply tonnage was the second highest on record.

By May, the Minister of State for Energy, Alick Buchanan Smith, had put the situation in perspective again in Houston: 'The era of easy money and easy profits in the oil industry is over. Large numbers of good people have lost their jobs and more may be likely to do so. That is a tragedy both for those affected and for the industry itself.' In December, the Government finally acted, relieving some of the pressures by repaying Advanced Petroleum Revenue Tax (APRT) earlier than planned, and freeing an extra £300 million for investment. The first indications of recovery came in the encouraging response to the tenth licensing round in February 1987. Eighty-four companies bid for nearly half the 127 blocks, and ultimately licences were awarded for fifty-one areas. The tide was turning in prices, too, as they reached $18 a barrel. But the most significant proof came with Shell's decision to develop its Kittiwake field and BP's to exploit the Miller field – both expected to cost about £1 billion each.

The crisis was over but the plummeting oil price had been a sobering experience for the oilmen, an intimation of mortality with lasting implications. Economist Martin Lovegrove said, 'As a whole, last year [1986] may have been a boon, in forcing it [the industry] to find cheaper ways of doing things.' The industry agreed with him. Speaking at a conference in 1997, Heinz Rothermund, president of UKOOA and managing director of Shell UK Exploration and Production said, 'The 1986 price crash, disruptive as it was at the time, was one of the best things that could ever have happened to us. For it imbued in us the mindset of continuous improvement. It helped us move into the current period – that of business management.' The high-spending days had gone forever, and even after the price returned to a more viable level, new rigid economies remained. A new and lasting financial regime was established. But the priority was still to maintain production, and the North Sea output for 1986 scarcely faltered. In 1988, the companies revived their exploration activities, renewed expansion and total oil-related employment began to climb again.

While there was to be another longer-lasting depression from 1998 to 2000, that 1986 slump is regarded as one of three epiphanies in the history of the UK industry. Another was Shell's public relations disaster in 1995 when they tried to decommission Brent Spar. The third came earlier, the 1988 Piper Alpha disaster; in human terms the most significant of the three. In the view of many offshore veterans, the new rigorous financial regime that sprang from the recession had a direct correlation with the tragedy a year later. 'When the price collapsed,' claimed OILC's Jake Molloy, 'the competitive tendering was such that it was a case of, "We can do it with fewer people and deliver a better product." And really at the end of the day, they couldn't. The controls, the supervision, the management, everything went out the window from the bottom right to the top and down again. There was no audit, no tracking. And you ended up with Piper.'

[8] Lessons from the Past

The crew on BP's jack-up drilling rig, *Sea Gem*, out in the southern North Sea, had just enjoyed a traditional festive lunch on Boxing Day, 1965. Derrickman Kevin Topham was lying in his bunk reading a book. 'This must have been when they started lowering the rig. It was like a lift going down in jerks. But this went down in one big jerk. I heard cracks. It was the legs. I thought, "This isn't the usual procedure," so I went out on deck. A few steps up and I was on the helicopter deck – it was absolute chaos. The 150-foot derrick had gone, snapped its mooring. The rig's legs were floating in the sea. The whole deck, where all the drill pipe was stored, was under water. I thought, "How am I going to get out of this lot?"'

It was the beginning of the end for the ill-fated *Sea Gem*, which eventually sank in the worst disaster in the short, hectic life of the UK offshore gas industry. Kevin survived. Fourteen others in the 32-man crew that Christmas died. The story of that first North Sea sector tragedy is worth recounting if only to ascertain if any important lessons in the practice of offshore safety or survival were learned and applied. Kevin doesn't believe they were. 'They said it would never happen again, because the *Sea Gem* design would never be repeated – "Lessons would be learned." The times I have heard that after so many disasters – ships, aeroplanes, rigs – it's like a long-playing record to me.' An official inquiry was held and the main change recommended was that offshore installations are required by law to have a stand-by vessel at all times. But because the tribunal had no statutory 'teeth' nor powers to suborn witnesses, it was hamstrung. Specific safety legislation for the offshore industry, the Mineral Workings (Offshore Installations) Act, which had prescriptive powers, didn't come into force until 1971.

One crucial aspect of what happened on *Sea Gem* which should have been marked for future reference was the total failure of the command structure leaving the blind human survival instinct to take over. Yet it was to happen again on another North Sea installation, twenty-three years later, and the cost in oilmen's lives was far greater. Piper Alpha was, of course, the

Plate 81.
The *Sea Gem* drilling rig spudding in 42 miles east of the Humber in June 1965, with an all-British crew. Six months later, on Boxing Day, she sank with the loss of fourteen men. *(BP plc)*

definitive watershed which forced a change in the safety culture. What exercises the present generation of oilmen, who see the toll in human life continuing, is whether or not safety offshore is again under pressure from the ever-present cost factor – as it was in the often callous early drive to develop the UK oil and gas industry.

Kevin Topham says that with *Sea Gem* everything had been done in a rush. 'The rig had been towed directly from the warm Persian Gulf and to the cold North Sea. They thought it was foolproof and that the 200-foot legs would never sink, but the changes in temperatures affected the metal. Divers hit one of the pieces of a leg with a hammer. It broke just like a milk bottle – shattered.' The operations engineer on *Sea Gem* was Jim Jenner. He had joined BP as a student apprentice, graduating in mechanical engineering. His first job was at Dass Island, in the Persian Gulf. When he returned to the UK, the first British rig was being assembled at Smith's Docks, in Middlesbrough. He was not impressed. '*Sea Gem* was essentially a flat jack-up construction barge, originally used for constructing the polders in Holland. At one end it had three legs on each side and at the other – the drilling end – it had two legs, all 8 feet in diameter. The jacking system was pneumatic with grips squeezed on to the legs and the rig simply climbed up and down. They had cut a slot in one end for the derrick and the rig floor. The drilling rig was old and it, the derrick and the power came from Trinidad, the pumps from Libya – just materials BP had lying around. The tanks and pipes were fabricated in

THE OILMEN

Smith's Dock. There was nothing in the hull. So it wasn't purpose built.' Like everyone else, Jim said he didn't know anything about its capabilities. 'It was a stopgap to go out into depths of between 90 and 100 feet. In my ignorance, I had accepted enough studies had been done to determine if it was viable. When I went out on it I was not terribly impressed, but I went out two or three times, and it drilled the well perfectly satisfactorily.'

Kevin, who had been in the Army for six years, had worked onshore for the Anglo-Iranian Oil Company at Eakring until the wells were depleted in 1965. He and most of his crewmates joined the *Sea Gem* in the summer of 1965 and began drilling in BP's freshly acquired licensing blocks. He was on board when they hit gas in what was to be the prolific West Sole field. 'It was the first hole we drilled. We didn't get particularly excited; we had struck oil and gas in places before. But it turned out it to be a huge strike – thirty years later it was still producing. So we had capped the well off and the Americans had begun to lower the platform, but it went wrong. Two legs snapped and the rig went to 45 degrees.' He relived that terrifying afternoon. 'When I went out on deck it was freezing cold, no rain, but a wind was blowing up. I went back for my lifejacket and a chap called Derek Stringfellow fastened it for me. We went on to the helicopter deck. There were fifteen of us. Then I realised no chopper could land. There was no Mayday, the radio cabin was the first thing into the sea. I decided to go down the steps into the water. When the rig eventually turned over people who stayed on the heli-deck went with it. If they had all gone down, more would have been saved, but they had gone to the highest point, which you were supposed to do. Two of us grabbed an inflatable dinghy at the end of the rig and it floated. We were just about to tie it to a handrail, which was under water, when the end of the rig dropped further and the waves flattened the float. We lost that one. Then we had to walk the handrail like a tightrope and we were getting really desperate. We couldn't see how we were going to get out. Eventually we got a second Carley float going and we got sixteen on to it – all that was left of the crew. We were in a helluva state, by then. The seas were getting up and there were no standby vessels in those days. We saw an empty lifeboat sailing merrily along and the radio operator, who was a champion swimmer, decided to go for it. I very nearly followed him, but he was younger and I didn't think I would make it in those waves. He pulled himself into the lifeboat. But by the time we were rescued, he was dead – frozen to death. I am damn glad I didn't follow him.'

Kevin said there had been no leadership on the float. 'We had never been in that scenario before. The only safety drill I can remember – if you could call it one – was in a hangar at Tetney Lock in a dinghy with two paddles –

Plate 82.
Kevin Topham (right), survivor of the *Sea Gem* disaster, takes a trip back over the North Sea. He now runs a small museum in Duke's Wood, Eakring dedicated to oil and gas memorabilia and archive pictures.

on dry land. Bit of a farce, but we never thought safety was particularly necessary. Certainly someone should have seen to it. We had no idea how we were going to be rescued, because no one had seen us, until this little cargo ship had come by. One of the crewmen saw the derrick and sent out a Mayday.' The cargo ship was the *Baltrover* and chief engineer, Leonard Woodhouse, said later that he couldn't take in what he was seeing – the rig tilting and obviously sinking. Kevin said, 'The ship tried to pick up what was left of us and that was the worst part. We had 20-foot waves and there were sixteen of us on a dinghy designed for twelve. Some were badly injured. I was okay. The cargo ship slung a net but it was knocking people against the sides. Two men were thrown into the water but they managed to pick them up. Then, as I went up in the net, I damaged my right leg. I was off six months and it's painful if I put pressure on it – even now. To my knowledge, not one man who survived put in a claim.'

Kevin never went offshore again. 'I have never had nightmares or flashbacks, but other survivors have had bad dreams. One man's wife told me he has the same nightmare every night.' Nor does he count the *Sea Gem* escape as among the very worst moments of his life. 'Six years in the Forces,

THE OILMEN

and half the time on bomb disposal, I have been in positions where things could have gone wrong. It was just part of life's experience.' There is a curious sequel. Kevin later worked for the Central Electricity Generating Board as a safety officer and one day he saw a welder using cutting gear on a pipe without a facemask. 'So I said, "Hoy, that's not on. You could lose an eye in a second." He looked at me, "I am on bonus." I said, "To me that is the biggest detriment there is – go and get protective gear." He said, "The store's 14 miles away." But I insisted. As he was walking away, he said, "You people talk about safety. You want to get on a North Sea oil rig and see what safety is about."' Twenty-three years on in a mature industry, Kevin said he had been amazed when he read about the lack of coordination, the lack of initiative, on Piper Alpha. 'Once again people went where they had been told to go and they perished. To me there is only one way off a blazing rig and that is down into the drink. There is no good waiting for leadership. It is all very nice on these safety drills but I know it doesn't always happen that way.'

In such a potentially dangerous environment, it is hardly surprising the offshore workforce is constantly exposed to serious injury and too often, to death. Even excluding major accidents involving rigs, platforms, vessels and helicopters, the North Sea has had the annual casualty list of an industrial battle zone, making it, after mining and construction, the third most dangerous industry in the UK. Piper Alpha was the wake-up call not only for the industry but also for the general public who were alerted shockingly to the real price of those fabulous oil revenues. Health and safety records, standards and policies now make bold headlines on the company and contractor mission statements to the public and to their shareholders and stakeholders. HSE is virtually a growth industry and while no one would question its necessity, its validity and transparency continues to be a matter for debate between workforce and management.

When the pressure was on and prohibitive costs were mounting in the early days, safety was well down the order of priorities. Yet the teenage Swede Lingard actually revelled in the risks. 'We worked together for years and I can't ever remember damaging anybody. We all had broken fingers or a trapped nail, and we used to do some really dangerous stuff, jumping about on the spider deck, jumping from beam to beam. Nobody wore safety belts in them days, or life jackets, except when you went down in baskets near the water level. But they were better days without a doubt, because they were pioneering days.'

One of his contemporaries, Joe Dobbs, said the key was teamwork. 'As for safety, we thought it couldn't be done any other way. We knew it was a risky business. But just before I left, it was getting on our nerves a bit

because we thought, "You aren't allowed to do anything now."' Joe remembers boat drill. 'They lowered this boat down off the davits. About 10 feet from the water they ran out of cable. It couldn't reach the sea. They had fixed the davits at high tide and when the tide was out there were a few red faces – we couldn't float it off.'

To young student Jim Jenner, who worked during his vacation on the land rigs at the Eakring, 'it was all pretty risky', but safety was something to be picked up very quickly. 'Nobody went out of their way to hurt you, but they did impress on you in a very practical way exactly what was going on – and it stuck. There is nothing like being scared shitless a couple of times to make you think about what you are doing. I have a horrible feeling people nowadays are pampered too much and they don't really pay enough attention, which is why they get hurt. Believe me, I think the work environment is a lot better than it was, much safer. But there is a happy medium. What we did do, we tended to look out for each other.'

A time-served boilermaker like Dave Robertson had been brought up on the shop floor, where he learned how to avoid accidents, unlike many of his workmates who had no background in heavy industry. 'These were things I learned as a laddie. But offshore there were so many accidents. These guys just didn't know where not to put your hands, where not to stand, what cranes do and don't do. When I went on the *Bluewater 3*, it had been drilling for quite a while and there were perfectly adequate crews. Then they brought out this new rig and the personnel guy took me aside and said, "I want you to go down to the Star and Garter and tell them there are jobs going." Sometimes ten new rigs arrived and there was nobody had any experience.'

In the early days it was planned to clear out the drilling area on the Forties platforms when the drilling programme went into decline and fit reciprocated gas compressors in the vacant area, next to the accommodation. Jim Souter was a BP OIM. 'We said, "We don't like this and we certainly don't want reciprocal compressors there. They can go up forward." Eventually they were mounted as far away from the accommodation as possible. Now, I don't know if we influenced them, but the fact they changed reflects much credit on them as it probably cost a lot more money. Otherwise we would have had a very similar platform design to Piper Alpha.'

One of Jim Jenner's early recollections of the men on the *Sea Gem* was that they had no appropriate clothing. 'There were no sea boots with toe caps, seamen's sweaters, donkey jackets, that sort of thing. The proper protective clothing did exist but hadn't found its way to the UK. It was a miserable experience out there and it took a long time before the industry got not only its safety act together but also got the guys properly dressed to do the job.'

THE OILMEN

Another issue centred on some debatable installation design. The West Sole A platform was long and narrow with two huge duplex mud pumps installed crosswise. The accommodation was on top of the engine house control room, three flights of stairs up. Jim Jenner was out at the beginning. 'The engineering people and the constructors were on a walk-round. We weren't actually drilling. We went for lunch and they served the soup. Then the drilling started and so did the pumps. Well, the soup went everywhere. The momentum and the reciprocation just made the platform sway from side to side. The chief engineer said, "What's going on?" I said, "They started the pumps," and he said, "Stop, stop." Eventually they cut the deck between the pumps and the accommodation otherwise the whole lot would have shaken itself to bits.'

One constant fear offshore is the prospect of falling overboard. Joe Dobbs's brother Brian was on *Sea Quest* while Joe was training in Yorkshire. 'Somebody came up to me – he was laughing his head off actually – and said, "Your brother has fallen off the basket into the North Sea." I didn't find it funny.' Brian and a mate, George Pemberton, had gone over the side in a basket to replace a mooring rope, damaged in a storm the previous night in October 1974. There were 12-foot waves hammering against the semi and the basket was swinging like a pendulum. Brian tried to cast a rope to tether the basket when his life jacket caught on the safety bar and opened it. Before George could stop him, he plunged 90 feet to the sea below. Brian said, 'I thought this was curtains, but when I surfaced I managed to grab a mooring line. I was in the water for about three minutes.' Had Brian gone under the rig they would never have got him out, but George signalled to the crane driver, who lowered the basket, and Brian managed to swim alongside. Three times George tried to pull his friend in but each time the waves kept them apart. At the fourth attempt George was himself dragged into the sea, but seconds later both men were washed back into the basket. George's bravery was recognised by BP and he was awarded a gold watch. Forever after, according to Little Joe, he was known as "Seiko George".'

The cranes were themselves highly vulnerable. Dave Robertson witnessed a high casualty rate in his days offshore. 'The drivers were the biggest risk group. A lot of cranes went over the side with the men in them. What happened was that in rough seas the rig is going up and down and the supply boat is going up and down. But the boat goes quicker. The crane drops its rope to pick up a container and the boat comes up, so there is loose rope. If it catches under a bollard then the boat drops. The cable doesn't break – but the crane goes over the side. There were two on one trip I was on in 1974. I also saw a jib come off *Bluewater 3*, but nobody was injured. They were very lucky.'

In 1982, Piper Alpha was attached by a bridge to an accommodation

Plate 83.
A crew wrestling to attach huge hawsers to the massive hooks of a crane – a graphic illustration of the size and scale of the machinery on board the rigs. Cranework was a frequent source of offshore accidents. *(BP plc)*

THE OILMEN

barge called the *Tharos*. The rigging broke and the bridge collapsed. Rov technician Derek Stewart said, 'Three men walking across were lost. We were scrambled with the early Comex ROV. We found the bodies lying 300 feet down with their hands upright. The Board of Trade inspector wanted to know how it had happened. We searched the seabed and found fresh gleaming wire rope. Occidental said it wasn't theirs, but it was. They got fined a paltry sum – £10,000. That rigging was just bad.'

One Christmas morning, when John Selbie was working on *Stadrill*, he and another roughneck, Gregor McKay, had just finished night shift. 'It was rough seas and a dirty day. We were watching a supply boat working alongside. There were two or three guys on the boat and suddenly this big lump of water came over the stern. The guys ran up forrard. One was swept off his feet and broke his leg. This other guy ducked between two containers, but the water lifted him like a cork right over the side. We all threw rings to him but he was a good bit away. They were huge seas, but he was a strong swimmer. Then we saw his arm going inside a ring. Now you are trained to lean on the ring and put your arms through, so that the ring is below you. But this guy was too tired to get it over his head. The supply boat went full ahead and burst both mooring ropes, bang, bang. They found him about an hour and a quarter later, but he was dead. Ten minutes in the water and he became unconscious through hypothermia. Because he had put his arm through the ring, when he lost consciousness, his head went down. He was still holding the ring but he had drowned.'

Like Brian Dobbs' experience, there were many narrow escapes in the North Sea and most went unrecorded. Schlumberger engineer Neil Ferguson was watching two men in a work basket. 'The guy working the crane put the basket down on the deck well, near the Rolls-Royce power generator, which had a huge exhaust – I don't know how many degrees of heat. He slung them over, but by the time he realised what was going on they had been over the top of the exhaust for a second. When he got them down, the heat had scorched all their facial hair; eyebrows gone. One guy's beard had burned off. The crane driver ended up getting fired.' Former merchant seaman Ian Sutherland worked on the early supply boats, which were always a high-risk. When he moved on to rigs, he found that safety was even more cavalier. 'I have seen them sending down pipes which were missing safety pins for the lugs that used to go round them. I shouted, "Just hold on." "No," they said, "we can't stop work for that." And a couple of tons of pipe would come hurtling down this chute. That would never have happened on a merchant boat. It would have taken, what, two minutes to fix – just nothing. To me it just wasn't worth the risk.'

Plate 84.
A precarious welding job for men who are secured by safety ropes. In the early days, according to some workers, risks were taken on jobs which should have required safety harnesses. *(BP plc)*

THE OILMEN

A simple example of the kind of changes in safety provisions introduced in the past few years came from John Nielsen. 'When we built Ninian North, the scaffolders didn't have to wear harnesses. They slid down a 20-foot pole, and if they didn't hold on they slid right into the sea. When Chevron went for full-time eye protection, there was hell to pay. People didn't want to wear it. Drillers said if they looked upwards to the monkey board into the sun, they couldn't see what was happening. But you just have to use common sense, so I would say, "Push your glasses up and have a look." Obviously, we learned a lot more as time went on.'

Sometimes it wasn't always possible to wear safety equipment. Mike Waller of Shell began his career on *Staflo*. 'On occasions somebody had to shimmy along this 12- or 15-inch pipe, 50 feet above the deck and knock a sliding joint along and tighten it up. No safety belts and no harness because there was nowhere to hang one. It was just a thing that happened. Something you had to do to get the job done. A lot of it was our ignorance, but we were helluva aware of our circumstances, I must admit that – you were tuned in for it.' In the 1970s, Derek Stewart, who worked as a diving technician, found safety provisions simply weren't available. 'Primarily, you looked after your own safety. Nowadays, you have to have the paperwork. Then, you didn't have a certificate and you didn't even need a medical. I have a medical once a year now and an RGIT safety and survival course every four years. You didn't have any of that, it was all word of mouth. We were so busy, so it was easy for anyone to join the industry.' In fact the first formal training courses were fully underway by the time Derek was travelling out to the oil and gas fields.

A series of industry accident statistics shows a steady spate of fatalities and serious injuries from the mid-1970s to the 1980s averaging just over fifty a year. These statistics from the Energy Department did not cover pipelaying or foreign registered vessels. The Trade Department were informed of accidents on UK supply vessels, but the official annual energy report said reliable information 'was not available'. Because of the continued rate of accidents, in 1972, Shell Expro and their main drilling contractor, SEDCO, realised formal training was needed. They approached Dr Peter Clarke, the new principal of Robert Gordon Institute of Technology (RGIT) in Aberdeen. The internationally known School of Navigation was chosen to run appropriate courses. Dr Clarke knew all about the industry. During the war, between university and his first job at ICI, he worked at Eakring and he had heard about the American drillers' top-secret operation. He came north in 1970 just as BP were announcing plans for Forties and he saw opportunities opening up in the new industry for education and training. Soon RGIT were presenting higher national certificate courses in oil technology and, in particular, drilling. The first survival course was part-time using existing staff. When demand increased, in 1974 the first full-time staff member was appointed, Joe Cross, former British Navy combat survival officer in the Arctic, and the UK's top expert in marine and air survival techniques. The development of Joe's survival and safety courses – and the rest of RGIT's considerable contribution to the offshore industry – will be described in a second book which will show how North Sea oil impacted onshore. Suffice to say that thousands of oilmen underwent the much-copied life-saving training, which included a unique helicopter underwater escape system, and totally revolutionised safety in the industry.

There was no discernible health and safety legislation at that time, and as Joe says, 'Safety and survival at that time was non-existent except what was done by the companies themselves. While they may have been doing lifeboat drills, there was no formal training in survival, using life jackets, life rafts, enclosed lifeboats and in helicopter safety.' Like other pieces of the industry's development jigsaw, RGIT was just as innovative (in the oilman's shorthand the course was simply called the 'RGIT'). In 1987 UKOOA's code of practice on safety and survival training was supported by the Offshore Petroleum Industry Training Organisation (OPITO), which validates the RGIT courses, incorporating firefighting and ranging over two to four and a half days. Currently no one can travel offshore without the requisite training certificates. Joe has retired and the centre is independent of the institute, which is now incorporated as Robert Gordon University. As RGIT Montrose, it is part of the training division of Petrofac and still the world leader in survival and safety training.

Plate 85. RGIT Survival expert Joe Cross, with the centre's chief training instructor, John Feather. The team created a standard of oil industry safety courses which became internationally renowned and much copied. *(BP plc)*

Plate 86.

The moments in RGIT safety training many oilmen dread – the realistic training in escaping underwater from a ditched helicopter. The machine and the techniques were devised at the centre. *(CNR International)*

But the real acid test is when oilmen are forced to apply what they have learned and survive. One grateful client was a helicopter pilot who ditched in the sea after a lightning strike disabled the rotor. He and his co-pilot and six oil workers escaped unhurt. At a conference on offshore safety, Captain Cedric Roberts described the emergency evacuation as a textbook exercise, but what made the escape more remarkable was that the lightning had also knocked out the helicopter's public address system. 'We thought the passengers were responding to our instructions, but we were amazed to find they hadn't heard a thing. Because of their training, they were running on automatic pilot, if you like.'

Another who returned to thank Joe and his team was Eric Morrans, who as a young man of twenty was one of two survivors of the industry's worst helicopter crash in November 1986. Two minutes from Shetland's Sumburgh Airport, a giant British International Helicopters Chinook, returning from Shell's Brent field with forty-seven men on board, suddenly plunged into the sea. Only Eric and the pilot, Captain Pushp Vaid, escaped; the others perished. The two survivors were fortunate. The crash was seen by chance by the airport's rescue helicopter on a routine training mission and they were picked up within 15 minutes. An inquiry later revealed the tragedy had been caused by corrosion in a modified gearbox. Shell immediately withdrew the Chinooks, which had superseded the costly fixed-wing and helicopter flights out to the Brent platforms. They never returned.

That tragedy happened at the lowest point in the worst financial recession in the brief history of the province, and with companies slashing costs, the offshore workforce feared for the role of safety in the order of priorities. There had also been a catalogue of other disasters. Since *Sea Gem*, the toll included: two men dead in an explosion on Cormorant Alpha in 1980; in 1984, three people burned to death on Brent Bravo; six killed in a blow-out on the Auk Alpha platform; and two years after the Chinook tragedy, two men died under a falling crane on a drilling rig. These were the headline-grabbing accidents, but when single fatalities and injuries are factored in, the figure is far more alarming. According to Charles Woolfson, John Foster and Matthias Beck in their analysis of capital and labour in the oil industry, *Paying the Piper* (1996),* the annual average for combined incidents between 1975 and 1985, was around sixty. In 1985/86, the figure was ninety. Excluding the Piper Alpha deaths and calculating on a three-year average, they claimed there had been a marked deterioration in safety conditions from that year and on beyond the disaster.

Protective legislation had been ostensibly in place since 1974 in the Health and Safety at Work Act, an extension of the onshore regulations, although unlike the mines there was no role for trade unions to become involved. The HSE were also subject to a complicated agreement, giving the Petroleum Engineering Division of the Department of Energy control over offshore occupational safety – a conflict of Whitehall departments opening the way to virtual self-regulation. Safety committees recommended in the 1979 report on offshore safety by Dr J.H. Burgoyne were sparse and at the discretion of the operators. There was also a severe lack of the new HSE inspectors. UKOOA, who, with the Government and the unions, were members of the Health and Safety Commission's oil industry committee, indicated they were satisfied with the Department of Energy as the regulatory authority and saw no need for HSE involvement. Oil workers' representatives were unhappy, however, at what they saw as an incompatibility of responsibilities, with the same department promoting the business side of the industry and also policing the safety regime.

---

* Woolfson, C., Foster, J., and Beck, M., *Paying for the Piper. Capital and Labour in Britain's Offshore Oil Industry* (London, 1996).

THE OILMEN

OILC founder Ronnie McDonald, who was still working offshore said, 'When the price collapsed it was "press the panic button". Drilling, exploration, appraisal all collapsed, maintenance programmes were shelved and the inevitable happened. Two or three years of cutbacks and Piper Alpha was guaranteed. On their own these wouldn't have caused the disaster – the unique uncertainties and instabilities – but the regulatory regime was deeply flawed, set up to ensure the line of least resistance for the operators. They wanted a greenfield environment, not just union-free but also free from the unwelcome interest of the regulators. That is basically what they were given.'

As a supply boat skipper and then as the procurator fiscal who reviewed oil incidents, lawyer Ray Craig is in a unique position to study the safety situation of that time. 'What you are still seeing are the same deaths and injuries occurring from the same basic mistakes. You can talk about risk assessment and you can talk about safe systems. At the end of the day, a guy is being pulled through the rotary table or whatever. Call it the need for command, control, supervision, or training – you name it – the legislation just wasn't strong enough before Piper, it was a damn shambles – the Department of Energy just couldn't cope, with only nine inspectors for the whole of the North Sea.'

The chairman of RGIT Montrose, Mel Keenan, worked at BP's refinery at Grangemouth, before becoming a union adviser. He later was employed in the safety sector by Elf. 'There were a number of concerns offshore at that time. One was obviously fire. In 1978, the oil companies got together and put up about £1.25 million to establish a fire school at Montrose and everybody who went offshore was trained at firefighting. There was certainly a lot more that could have been done, but nobody had the foresight or the vision, neither unions nor employers. No one asked what would happen if there was a tragic event or a fire or explosion and there had to be an evacuation. It seems strange now but there were no established procedures either for evacuating skyscrapers, factories or cinemas.' He said that in the 1980s, safety procedures existed but were not enforced as rigorously as now. 'All too often they learned through having accidents. In the early days the people on the rigs were very experienced. They had vivid memories of accidents, so they could run the rigs pretty accident-free. In later years, when people started leaving the oil industry, there were few experienced staff people left.' Neil Ferguson, who was working for Schlumberger, said that before Piper Alpha there was a safety regime. 'Each company had a plan, but whether that actually filtered down to the workforce is another matter. Like everything else, it cost money. When things got busy, it was the first thing that took a back seat.'

Plate 87.
Shell's Shearwater field in the central North Seas – its gas condensate is pumped to Bacton in East Anglia by the SEAL pipeline, while the oil links into the Forties transport system.
(Shell International Ltd)

By the end of the 1980s, the industry was generating huge streams of revenues. Exploration continued, with 129 wells drilled up to 1988 from which twelve significant discoveries were made, among them Harding, Saltire and Shearwater. A total of 560 exploration, appraisal and development holes were drilled, while the thirty-six offshore fields produced a total of 114.4 million tonnes of oil, and twenty-four gas fields had an output of 42 billion cubic metres. Capital expenditure – estimated at £2 billion – was 11 per cent of the UK's total industrial investment. Total oil revenues were £7.2 billion, while the gas fields brought in £2.1 billion. Employment was creeping back up to a total of 29,300 offshore, 90 per cent UK nationals. As further confirmation of commitment, eighty-four companies applied for 125 applications for the 212 blocks on offer at the Eleventh Round in the summer of 1988. To service the expansion, onshore in Aberdeen, Peterhead, Montrose, Invergordon, Shetland and Orkney, a huge array of bases, offices, workshops, and research centres had sprung into existence; the refurbished Aberdeen Airport and expanded Aberdeen Harbour were breaking all records for air and sea passenger and freight transport, and the construction yards at Nigg, Ardersier, Skye, and Methil were still launching new production platforms and modules. An international industry had come of age as a major Scottish enterprise.

THE OILMEN

Plate 88.
The hi-tech nerve centre of the modern offshore installations: inside the control room of BP's Schiehallion floating production system, west of Shetland. *(BP plc)*

Plate 89.
In contrast to the Schiehallion control room, one of the early wire-logging rooms before the development of computers. *(Geoservices)*

Plate 90.
One of the original offshore fire fighting vessels specially built for work in the North Sea – BP's *Iolair* with her water jets in full array. *(BP plc)*

The offshore industry stabilised as a series of marine 'industrial estates' of busy factories, each with an employment complement comparable to any medium-sized company onshore. The complex pipeline systems were feeding the scattered terminals round the coast and the islands. The contracting sector now overwhelmingly populated the offshore installations, save for a core of operator staff. The North Sea was the world leader in offshore innovation as gradually families of new technologies were being brought into play. Subsea completions with lengthy tiebacks, high pressure, high temperature wells, computerised drilling, new downhole tools and injection systems, were all revolutionising the oilfields. On the survey side, seismological testing was more accurate through new forms of dimensional imaging. Like sizeable onshore power stations, the platform nerve centres and processing plants were fully automated, the short-sleeved shirts of the engineers outnumbering the muddy overalls of the floormen. Computerisation was transforming the dangerous and physically demanding toil on the drill floors. The modern platforms and installations were now models of accommodation and social facilities. The supporting airborne and seaborne armies were also keeping pace: new generations of helicopters, with refined navigational aids, pre-flight instructions and survival suits; giant

THE OILMEN

custom-built supply, standby and emergency response vessels, complete with helipads and bristling with electronics, were replacing the converted trawlers and coasters; specialised diving support ships had appeared, and more sophisticated remotely operated vehicles took to the water; and firefighting and pollution-control vessels had become an integral part of the blossoming oil fleet. On all fronts the North Sea was advancing ahead of the rest of the world. Yet while modern approved survival training had also become a fixture, safety, as the oilmen have already claimed, was lagging behind all the other advances.

A number of companies would have disagreed with that appraisal as far as their own operations were concerned. The American company Conoco was generally recognised as having a good safety record. Bill Schmoe was in charge of the UK upstream activities. 'As a corporation, we were very safety conscious and led the North Sea almost every year in safety statistics. That stemmed from L.S. McCollum, the man who led the company's international drive. He was a stickler for safety, and when the top man is that way, everybody else is, right down the line. That is how I was brought up in the oil business.' But that situation was mostly self-regulation, more company-led than Government-sponsored. The facts were that no palpably rigorous regime was in place and there was limited enforcement because of the fragmented legislative safety structure. That macho culture was about to change cataclysmically and tragically on the night of 6 July 1988.

Who Listened to the Piper? [9]

In the Moray Firth licensing sector of the North Sea, in the fertile fields which the international power broker Armand Hammer was said to have acquired in a clandestine settlement with the Foreign Office, the Occidental platforms – Piper Alpha, its identical sister installation Claymore and the Tartan – had been pumping out copious amounts of oil and gas since the late 1970s. From the beginning, Piper had been recognised as the world's most prolific producer of oil from a single platform at 284,000 barrels per day. Because of the proximity of the fields – Piper was only 22 miles from Claymore – the three platforms were connected by a pipeline system. As the central installation, Piper acted as the receiving and pumping station for gas and oil which was then fed to the Flotta terminal.

Piper had a troubled history. In 1982, a bridge collapsed between the platform and the accommodation facility, the *Tharos*, and three men died. Following a prosecution two years later, Occidental and the contracting company, Strathclyde Process engineering, were fined a total of £15,000. Dave Robertson recalls another incident on the platform, previous to the disaster. 'The Piper was one of the few platforms that had a safety committee. My mate was there at the time. There had been an explosion in the gas compressor – same place as the big disaster – and a union representative came to settle the guys down. They had a meeting and he asked to see the results of the inquiry that had been held, but he was knocked back. The safety committee had been set up by the management and had no strength of its own. It was only there to ask the right questions.' An official report on the incident was never made public and Occidental were never prosecuted. Dave said, 'One of the guys who resigned over that incident was one of the ones who got killed in the Piper disaster, Bobby Adams, a very good friend of mine. But I knew an awful lot of the guys. I remember the first time I was on the Piper, a guy said to me, "See this rig – it will be the first platform on the moon."'

As the TGWU organiser, Mel Keenan had visited the platform on a number of occasions. 'All the talk that goes on now about safety and how it

Plate 91.
Occidental's Piper Alpha platform in the Moray Firth sector – producer of the most prolific daily oil output in the North Sea.
(Kenny Thomson/ John Greensmyth, Technip Offshore)

Plate 92.
Opposite. Piper Alpha caught in the red light of its flare – an eerily prophetic image of what was to come.
(James Fitzpatrick)

would have been much enhanced if the unions were involved – Piper Alpha was the only platform at that time that was covered by union agreement. I think this is one of those times when everybody had twenty–twenty vision in hindsight. Claymore is constructed in exactly the same way and it is still there today. Piper had problems. The equipment was manufactured by people with no experience of installations in the harsh environment of the North Sea, and there were difficulties with the sprinkler system, but I don't remember people saying it was a death trap or badly maintained. Occidental were operating in the same way as the other companies.' Ann Gillanders' husband, Ian, would have disagreed. He had actually worked on the construction of the installation at Ardersier. She remembers overhearing him telling a friend about the poor state of the platform and the presence of rust and corrosion. Ian was one of the ones who died.

Twenty–twenty hindsight or not, it is still a commonly held opinion that it was '*the* disaster waiting to happen'. A variety of reasons are cited, including cutbacks in maintenance during the oil price crises and the absence of a mandatory safety culture. Yet Ian Gillanders' friend, Bob Ballantyne, was annoyed with those who talked about it in this way. As an electrician, he

THE OILMEN

worked on many installations across the North Sea. 'Piper Alpha was no better and no worse than any other. Better, in fact, than some. But they were all the same – they all needed work done on them.' Bob, who has since died, was one of the sixty-two survivors of the tragedy.

Dave Allan, a wire line supervisor, was an Occidental employee who worked alternately on Piper and Claymore for about five years. 'Me and a colleague had come off the morning before from the floater. We went into the Skean Dhu for a drink and met guys who should have been on Piper but were bumped off the helicopter because they needed space for equipment. So we had a few pints and got these guys pie-eyed. They were going offshore later. That was the last we saw of them. I was actually in bed on Thursday morning when I heard this scream and that was the wife. It was on television. One of the guys I drank with that night had the first funeral in Aberdeen – Don Reid. I said to his wife, "I saw Don." She said, "I know, he phoned me that night at nine o' clock. He said, 'Hey, I've got a helluva hangover.' He was on permanent night shift so he wouldn't have been working. "Bloody Dave Allan and his mate. Thank Christ, I was going to bed."'

Piper Alpha was the worst disaster in the turbulent history of the international oil industry and the graphic pictures commandeered the headlines of the world's media. Across the regions of England, Wales and the central belt of Scotland, where the majority of the dead had come from, families grieved. But it is probable that the impact was most keenly felt in the area most identified with the industry, the north-east of Scotland, sadly all too familiar with the price the sea can exact in the lives of its fishermen and seamen. Oil was the north-east's industry and it was the north-east's workforce, no matter that many of the 167 who died stemmed from other parts of the United Kingdom. It marked the end of a romanticised dream of an industry only known to the uninitiated through the celluloid glamour of Hollywood. Suddenly, it had become shockingly and brutally real.

The cruel facts were revealed a year later during the thirteen-month public inquiry held by the eminent Scottish judge Lord Cullen, from the evidence of 260 survivors, eyewitnesses and experts. The tragedy was set in motion by the removal of a pressure safety valve from a pump during maintenance and its replacement with a blank flange. Crucially, Lord Cullen discovered, this fact was not communicated to the oncoming night shift. When the pump which boosted the gas condensate through the pipeline broke down, engineers started what they presumed was the relief pump. The condensate rushed through the flange and ignited in an explosion, disabling virtually all the emergency systems. The two other platforms in the system

Plate 93.
The headline that shocked the world – the first report on the world's worst-ever oil disaster in which 167 oil workers perished.
(Aberdeen Journals Ltd)

Press and Journal

SPECIAL LATE ABERDEEN EDITION

241st Year — THURSDAY JULY 7 1988 — No. 41,587 — 24p

193 OILMEN FEARED DEAD

5 a.m.

Inside story

Ship rams rocks

AN OILFIELD rescue ship rammed into rocks near Lerwick Harbour in fog last night.

The crew of 10 of the Seabased Intrepid took to a damaged lifeboat and reached Lerwick three miles away. The captain was taken off the vessel by Lerwick lifeboat.

The engine-room was 5ft. deep in water when they abandoned ship, and a crewman said it was lucky no one was injured.
— PAGE 12

Graduation stories, pictures — PAGES 9, 10 and 11

Act fast, Newton demands

HEALTH Minister Mr Tony Newton last night called for fast and effective action in Cleveland to clear up the mess left by the child sex-abuse scandal.
— PAGE 2

Travel firm fail

AFTER the financial collapse of an Aberdeen "bucket shop" travel operator, angry creditors were told their air ticket advance payments had probably been spent on alcohol.
— PAGE 13

On other pages:
Births, Marriages, Deaths 2
District news 3, 6
TV Guide 4
Andrew McKay 4
World news 5
Horoscopes 12
Cartoons 13
Farm Journal 15
Business 16, 17
Sea writer 17
Classified 17-22
Sport 23-24
Crossword 24

Piper Alpha platform blown apart

By DAVID STEELE and IAIN LUNDY

MORE THAN 190 people were missing feared dead last night in what could be the world's worst oilrig disaster.

A massive explosion blew apart the giant Piper Alpha platform, 120 miles north-east of Aberdeen. Early this morning only 39 people were accounted for of the 232 on board.

Rescuers last night described the platform as a "raging inferno" and coastguards spoke of bodies floating in the sea one-and-a-half miles away.

Many of the survivors were flown in to Aberdeen Royal Infirmary this morning with severe burns. Some were picked up from the water by a fleet of rescue helicopters which converged on the scene after the blast.

The 34,000-ton Piper Alpha is operated by Occidental. A spokesman at the RAF Search and Rescue at Pitreavie last night said the platform was "on fire from top to bottom".

The tragedy appears likely to prove far worse than the Alexander Kielland, the rig which capsized in the Norwegian sector of the North Sea in March, 1980, with the loss of 123 lives.

The explosion shattered the Piper Alpha at 9.31 p.m. and was followed by a raging fire. It is not known what caused the initial explosion, the third to hit oil installations in the past week.

Four RAF helicopters, an RAF Nimrod and the Sumburgh-based coastguard helicopter joined a flotilla of vessels in the fire-fighting and rescue operation. Some of the survivors were winched from the sea and others were picked up by passing boats.

One unconfirmed report this morning indicated that a vessel helping in the rescue had blown up.

A Naval task force of six vessels was this morning heading for the disaster scene, as was the minesweeper HMS Blackwater and the P&O ferry St Clair.

The multi-purpose vessel Tharos, which contains emergency medical facilities, treated many of the survivors before they were flown to Aberdeen. Others were treated at Occidental's nearby Claymore platform.

The full extent of the tragedy began to emerge early this morning when rescue services were able to account for only 39 survivors, leaving a total of 193 missing.

The Tharos, specially designed for such emergencies, has accommodation for 300.

The Piper platform was hit by an explosion previously, in 1984, when 55 men required treatment.

Last night's incident followed an explosion the previous night on the Shell-Esso Alpha platform further north, where oil production has been halted. In that case the explosion was in a gas-compressor. And BP are investigating the cause of an explosion in a gas-compressor at the Sullom Voe terminal in Shetland last Friday.

Pitreavie Search and Rescue Centre said non-essential personnel from the Tharos had been evacuated from the vessel to make way for casualties from the Piper Alpha.

Weather conditions on the field were said to be "reasonable" with visibility about five miles and a southerly wind about force four or five and waves about a metre high. The conditions should not hamper the rescue, said a London Weather Centre spokesman.

Caithness and Sutherland M.P. Mr Robert Maclennan last night said there would need to be a thorough inquiry into the incident.

"We need to find out what has gone wrong," said Mr Maclennan.

It is believed Energy Secretary Mr Cecil Parkinson will face demands for a statement about the incident in the House of Commons today.

● A major incident room was set up at Grampian Police headquarters in Aberdeen and relatives of those on board the Piper Alpha should call 0224 649606.

WEATHERGUIDE

Aberdeen, Moray Firth areas: Sunny periods in the morning, showers developing in the afternoon. Warm with light southerly winds.

FULL FORECAST — PAGE 13

Picture by SANDY McCOOK

SAFE ... a wife hugs her oilman husband after he was flown into Aberdeen with other survivors from the platform blast.

ANOTHER PICTURE — PAGE 12

THE Piper Alpha oil-production platform.

Sobbing woman greets survivor

By DAVID STEELE

IT WAS 3.20 this morning when the first helicopter arrived with the dawn at the helipad at Aberdeen Royal Infirmary.

Ambulances waited to rush the victims to the accident and emergency unit, a few hundred yards away, where doctors and nurses were ready to go to work on the injured.

A handful of shocked relatives were in the waiting crowd, which included a large Press and TV contingent.

Seven men walked from the Bristow Puma and two were carried out on stretchers. One of the latter was obviously been badly burned and there was a cry for resuscitation equipment as he was carried into the hospital.

As the men walked from an ambulance into the hospital building, a sobbing woman fell into the arms of one of them, obviously overcome to know he was among the survivors. They went into the hospital still clinging to each other.

The helicopter had no sooner disembarked its passengers than it quickly lifted off again, apparently heading back to the disaster scene.

More helicopters were expected at the hospital this morning.

Flt. Lt. HODGSON

SEA KING helicopter captain Flt. Lt. Steve Hodgson who was the first pilot sent to the scene of the disaster last night, has been in the front line for days.

He was on his way to a mountain rescue in the Cairngorms last night when the call came to Lossiemouth where he re-fuelled to fly to the Alpha platform and the gruelling task of searching for survivors.

Earlier this week he was at the centre of the drama in which premature Orkney baby Sam Harcus — born weeks early at his parents' wedding on Sunday — was rushed by lifeboat to Kirkwall where Flt. Lt. Hodgson had to put down in a field because the airport was fogbound.

Last Friday, he had to land, again in dense fog, on the Beach Esplanade in Aberdeen with an English diver who was suffering from the bends.

RIG BLAST LATEST

OCCIDENTAL confirmed only 40 out of 232 accounted for. Production from nearby Claymore and Tartan platforms shut down as a precautionary measure. Peter Morrison, Minister of State for Energy, holding Press conference at Oxy HQ at about 8a.m. Senior Oxy executives from London and California flying to Aberdeen. Oxy inquiry beginning immediately.

Serious accidents part of life on rigs

SERIOUS incidents aboard North Sea drilling and production platforms have been a part of life since the 1960s and the start of the offshore exploration boom.

Structural defects, explosions or the weather have all been to blame.

The first "blow out" was in April, 1977, when at least 20,000 tons of oil escaped into the sea from the oil-rig Bravo in the Norwegian Ekofisk Field run by Phillips Petroleum.

Texas oilman Mr Red Adair was brought to Britain to cap the fracture. It took Adair and his team five attempts to cap the escape, but not before a slick covered 1500 sq. miles.

The most serious incident was in 1980 when 123 crewmen aboard the accommodation platform Alexander Kielland died after she capsized in storms in the Ekofisk area.

A ruptured oil pipeline had caused an explosion which killed three men and seriously injured three others in the same block in 1976.

Three people died and 11 were injured when an explosion ripped the bottom out of a drilling barge at Teesport, near Middlesbrough, in November, 1965. A month later, 13 men were killed when the Sea Gem collapsed and sank in the North Sea.

Advice to pilots

BRITISH Airways have reviewed instructions to their pilots flying over the Gulf following the Iranian Airbus tragedy. BA chief executive Sir Colin Marshall said there were all sorts of precautions their pilots could take.

No stranger to accidents . . .

OCCIDENTAL'S Piper Field came on stream in 1976 and the 34,000-ton platform is no stranger to casualty.

Three men died in October, 1982, when they fell 70ft. from an access gangway on the platform into the North Sea.

And in 1984, 175 men were evacuated from the platform when it was rocked by another explosion. Fifty-five received hospital treatment at that time.

Piper Alpha can hold a maximum of 200 workers and the average daily production in 1987 was 167,290 barrels.

Occidental Petroleum are one of the major operators in the North Sea and have been run for 31 years by American multi-millionaire Dr Armand Hammer.

Earlier this year the company announced a new find five miles from the Piper Field. They are spending £30 million appraising the possibilities of another find in the more remote Birch Field, and won permission at the end of last year to develop the Chanter Field.

THE OILMEN

Plate 94.
The eminent Scottish judge, Lord Cullen, whose ground-breaking report radically changed the safety culture of the oil industry.

were too slow to shut off the flow of oil and gas to Piper, compounding the growing conflagration. There was a second massive explosion. The emergency deluge water-spray system failed and no one could reach the standby diesel pumps. The men in the accommodation module, positioned over the gas processing plant, had no chance as they waited for rescue in their living quarters, which ultimately plunged to the seabed. The sixty-two desperate men who survived the phenomenal heat and flames had to find any escape route they could across the scorching hot decks, some suffering from severe injuries, before leaping hundreds of feet into the sea. They were saved by the bravery of the crews of the little fast rescue craft launched from the standby boats, one of which was eventually blown to pieces. Only one crew member escaped; the rest, along with the oilmen they had sped back to rescue, perished. In contrast, the massive *Tharos* semi-submersible, specifically designed to respond to such emergencies with firefighting systems, was powerless, reverting to the role of a medical relay centre for the badly injured and traumatised survivors ferried in by helicopters. Like *Tharos*, the helicopters were unable to reach the stricken Piper Alpha in the extreme heat and impenetrable smoke.

Bob Ballantyne remembers those terrible hours as 'a surreal kind of Hell'. Bob had learned the trade of electrician in the Clyde shipyards and was a committed trade unionist. He had worked offshore for eleven years. At about ten o'clock on the evening of 7 July he was in the recreation room on the platform entertaining his mates with one of his stand-up comic routines. He remembered the laughter. Then there was a mighty explosion and the platform shook. 'There was total confusion. We rushed out into the

Plate 95.
In the cold light of day, a helicopter searches for crewmen. At the height of the disaster, the oilmen's aerial ferries were powerless to help in the confusion of smoke and heat. *(Bristow Helicopters)*

THE OILMEN

corridors. They were already full of smoke. I was with a bunch of guys, including Ian Gillanders.' They made for the galley where there were already about a hundred people.

Former RAF radio officer Mike Jennings first went offshore in 1983, and in 1984 started work as Piper Alpha's flight information officer. He knew about the platform's poor safety record and about an evacuation which had taken place three months before he went there. But he had no qualms about it and he enjoyed his job. When the first blast came, he had just finished a twelve hour shift and was in the cinema. 'From the thump and the actual uplift of the floor – it actually reverberated and bits and pieces fell down – you knew it was serious. The cinema was fairly packed at that time.'

Bob Ballantyne described the noise and the intense heat. 'Our faces were black with the smoke. The rubber in the survival suits was melting. My instinct was to go down, although that was against safety procedures. But there was total chaos. We decided to split up to look for a way out. I headed for one side; Ian and the rest went to the other. I never saw them again.' Bob and Ann Gillanders think his mates were taken off by the rescue craft that exploded. Ian is among those whose remains have never been found.

Mike said the cinema was in the accommodation block about six stories from the bottom level. 'I went up to my muster station, the radio room. But it was empty. Flames and smoke everywhere, tremendous heat – I could hear other vessels putting out Maydays for us. I put my survival gear on, then tried to get down to the lifeboats. But people were coming up saying, "There is no way down there."'

Plate 96.
One of the Piper Alpha survivors, Bob Ballantyne. The North Sea veteran radically altered his way of life after the tradgedy. Each day he makes a point of spending ten minutes just thinking about his escape and about his lost workmates. *(Aberdeen Journals Ltd)*

But Bob had reached the bottom level. 'The sea wasn't on fire, but it was like daylight with the flames. There was debris everywhere. It was too far to climb down – about 60 feet. Then I saw this rope. Somebody had left it tied to a rail. I couldn't believe my luck. I went down into the water. It was warm. I clung on to the base. Then there was another explosion and I could see the metal melting. So I swam for it. I was trying to shout, but I couldn't hear myself for the tremendous noise. The strange thing is one of the boat crew who rescued me told me later, "Above all that din, all we could hear was you shouting – clear as anything – and the language!"'

Mike had gone back into the dining room, the designated evacuation centre. 'The OIM was saying people would come to rescue us. But I knew no helicopter could land and other people told me the lifeboats hadn't been launched because there were no coswains. I don't know why they stayed and I went. But I had been outside. I had seen the flames coming up the sides of the windows – it was like a pot sitting on top of a gas stove. The ones who stayed there were lost.'

[188]

Bob Ballantyne regularly thought about his ordeal in the seas around the blazing platform. 'I usually spend about ten minutes every day just thinking about it all. I was told that would be good for me. I have asked philosophers why they think I survived, and all they could come up with was "pure luck". I think it was love. I was about 45 minutes in the water and all I could see was my daughter's face – and my partner, Pat. They were what kept me going.'

Mike eventually reached the pipe deck. 'Then I saw somebody walk along some pipes and jump over the side – about a hundred feet. I thought, if he can do it, so can I. It was the only way out. I stood looking down and somebody just pushed me. I don't who he was or why he did it. I don't remember hitting the water. When I came to the surface there was fire everywhere. A piece of partition floated past, so I pulled myself on to it.'

Bob was eventually helicoptered back to Aberdeen Royal Infirmary. Remarkably, he was not badly injured. 'I had burns to my wrists where the rubber cuffs on my survival suit had melted, and my face was like a bad case of sunburn. I was one of the lucky ones. Some of the others . . .' The electrician never went offshore again, changing his lifestyle completely. He gained an honours degree in cultural history at Aberdeen and taught people with special education needs.

Mike said he was more or less OK. 'Just small burns, my hands and my head from stuff falling on it – and my shoes had burned.' He was picked up by the *Silver Pit*. 'They pulled me up the scramble nets and it was hard going, but that was all the equipment they had. Again it was a case of cutting costs and standbys were always the first to be cut back. I was in hospital for four days. They put me into intensive care inititally because they thought I was suffering from secondary drowning, and there were people with horrific burns. I felt guilty. I had got away, but seeing those guys I thought, "I am lucky." '

The eighth of July was a long, agonising day for Ann Gillanders and her family at their home in Nairn. She had woken her daughter and then gone back to her room in time to hear a report on the radio about an explosion on a platform. Piper Alpha. She knew her husband Ian was out there. He had been working there since November 1986. 'I just couldn't believe it. There was that vain hope that maybe he had been transferred. I started phoning. I phoned his company, the Wood Group, and couldn't get anybody. I tried Occidental. I was just phoning and phoning for hours. Finally I got somebody at the Wood Group. Just that hope, you know, but she said, "I am sorry, I've checked." They didn't have any further information.' She spent the rest of the day trying to find out more as the family gathered around her. 'It

Plate 97. Ann Gillanders pictured when she was one of the leaders of the Piper survivors and families group. The body of her husband Ian has never been recovered. *(Aberdeen Journals Ltd)*

THE OILMEN

was six o'clock when a policeman came to the door. I just heard "missing, presumed dead". Missing – but I think I knew in my heart. The policeman said, "You realise there is not really any hope?" I said, "Yes." He was maybe thinking, better that way rather than hoping when there was no hope.'

While Mike was in hospital in Aberdeen he had a number of visitors – but no one from his company, East Anglia Electronics, or from Occidental. 'I was only in hospital four days, long enough, I remember, for a visit from Prince Charles and Princess Diana, and Mrs Thatcher. When I came home I went to the funeral of the communications technician who had been with me.' Then incredibly, after only three weeks, he decided to go back to work. 'I just felt I had to do it. If I hadn't done it then, I would never have done it. I went on to Claymore in my old job. I had to occupy my mind and not think about it.' Apart from the guilt that Bob Ballantyne also talked about, Mike felt anger. 'I was generally Mr Angry and I was not normally like that. My wife said I was a changed person and maintains I still haven't got over it yet. There was all this clamour in the press for a flypast over the site. So I was angry with them and I was also angry with Occidental because they were trying to prevent it. I even rang up the general manager and asked him, "What have you got to hide?"'

Ann Gillanders was beginning to feel the same anger. 'As far as the Wood Group was concerned, if there was any information they usually got in touch. As far as Occidental were concerned, apart from their own workforce, they more or less passed the matter on to the contractors. You did hear from them occasionally. But what I felt was that press releases only came out every now and then when they were pre-empting criticism – a kind of damage limitation.'

The Cullen inquiry brought the survivors and the victims' families together. Ann said she had to go. 'I had to hear what had happened, because I hadn't been there and I hadn't been with him when he died. Not everybody felt the same way, but I was still hoping his would be one of the bodies found. Charlie McLaughlin from Glasgow, his body was found, and he had been in the group with Ian and Bob Ballantyne, who phoned me a couple of days afterwards. I really appreciated that. It must have taken courage. I knew sometimes the ones who have lost people can resent survivors. I didn't feel that way. I just said to him, "I am so glad you got off."'

Grampian Region set up an outreach group and helpline for the families and survivors, which operated for more than a year. The families used to meet as a group to discuss the latest developments. 'We felt we were powerless. If it was anywhere else, such as a factory onshore and the company was considered negligent, other agencies would take over and the owners wouldn't get near it. Because it was offshore it was Occidental who had the final say.'

Occidental sent Mike Jennings to various psychiatrists and psychologists. 'I am not sure they did that for everybody. It didn't do me any good. I don't know whether they were trying to avoid paying me compensation.' He also went to some of the survivors' meetings. 'It actually helped to see other people with the same kind of problems as me. I was lucky I came away reasonably sane and reasonably unscarred. But I would much rather never have been there in the first place.' A great deal of work was carried out in Aberdeen in treating survivors and their families for what was then the little-known disorder post-traumatic stress. But at each stage – the retrieval of some of the victims' remains, the inquiry and the ultimate report – the ordeal was prolonged and agonising. Thirty men were never found. The ashes of one unidentified victim are buried – at the families' request – at the base of the memorial statue. The crew of the *Silver Pit* were later honoured for their exceptional bravery in their repeated attempts to pick up survivors.

As the disaster was unfolding, Jane Stirling was one of the three structural engineers called in by Occidental. 'We sat through the night getting out drawings for the emergency response room. And people kept coming out with things like; "Module D has gone." "Gone? What do you mean gone?" We were analysing structures without realising how horrific it all was. The next morning, when I went home in a taxi, the driver threw me the *P&J* [*Press and Journal*] with the picture of the remains on the front. Now I knew this platform inside out, back to front, and I couldn't get my head round what I was seeing. Obviously lots of people I knew very well died that night, so there was a horrible feeling. It affected everybody and many of us had counselling afterwards, but they always say the best way through something like that is to keep busy. Well, were we kept busy!'

Procurator Fiscal Ray Craig was responsible in law for the bodies of the victims. 'The fiscal has to ascertain who died and what he died of. My first experience of Piper Alpha was literally hearing of it that evening and next day going up to the hospital to show the flag, basically, because I knew the police would report to us. I saw the first seventeen bodies ashore and then a number of the rest. I had no involvement with the witnesses who were actually at the scene; it was peculiarly silly of the fiscal service to waste that experience.'

The disaster touched people in the industry in many ways. Mel Keenan, who had union members on the rescue craft that was blown up, reached a stage where he had to get away, so he found a job with the Banking Insurance and Finance Union and went back to the Central Belt. 'I wasn't

THE OILMEN

feeling too good about the world. I had visited all twenty-five families of my members who died and that was one of the worst weeks of my life. But I felt it had to be done, because they were asking questions. It was my duty to do it and also to represent them, because every single member of the TGWU involved was represented by the union in court.'

Later, Ray Craig was drafted in to investigate one strand of the disaster for the inquiry, but he now questions the apparent lack of direction he was given. 'My role was to interview the inspector for what was then the Department of Energy. He had inspected the platform the year preceding the incident. In my view he had absolutely no responsibility for what went wrong. But the inquiry was a funny situation. They talked about Lord Cullen and the esteem he was held in – my doubt is I don't think judges are necessarily the best people to do the investigative process. My own experience was we got no guidance into what was being looked for. It was my determination of what I thought it ought to be. I don't think you can run an inquiry on that basis. Having said that, nobody disputed the work of the Cullen Inquiry. It was just a strange way to work it. It meant, perhaps, they themselves didn't know what to look for outside particular areas.'

There was a second stage to Lord Cullen's report in which he analysed in forensic detail the full gamut of the existing safety and survival systems in the North Sea, and then produced a series of seminal recommendations. To many, Piper Alpha represented all that was wrong with the industry and it was hoped that Cullen's comprehensive findings would be the catalyst for a new, safety-conscious culture that would prevent such tragedies happening again. Part of the main thrust of the 106 recommendations, which were totally accepted by the Government, transferred safety from the Department of Energy, which was criticised, to the Health and Safety Executive. The HSE later created the offshore safety division based in Aberdeen. Other recommendations included the production by the operator of a safety case detailing safe management systems and identification of hazards; the introduction of temporary safe refuges, and the 'support and encouragement' of safety committees and safety representatives' to be regulated by the HSE. Lord Cullen said the onus was on each management to change its approach to safety. 'It is essential to create a corporate atmosphere or culture in which safety is understood to be and is accepted as the number one priority.'

Occidental eventually did pay compensation to the survivors and their families, then sold its assets, including the Piper network, and left the North Sea. Mike Jennings did not receive any major compensation because he had returned to work. After the Cullen Inquiry, the survivors and the families continued to fight for some form of retribution, but although severely

Plate 98.
Striking images of the Piper Alpha memorial in Hazelhead Park rose garden in Aberdeen. (Photographs: Kate Sutherland. Montage: Greybardesign)

castigated, Occidental never faced any legal action. Some of the Piper survival group joined victims and families involved in other major disasters, such as Lockerbie and the King's Cross rail fire, to campaign for a legal charge of corporate manslaughter to be brought in for cases such as Piper Alpha. The concept is still being considered by the authorities.

Ian Gillanders' company suffered the heaviest losses. Forty-seven of the 167 who died worked for the Wood Group. The chairman, Sir Ian Wood, said later, 'Piper Alpha has left me deeply scarred. It was a dreadful scar on the North-east and on the Wood Group. You have to live through it to understand. You finish up with huge amounts of guilt and concerns that you have to come to terms with over a period of time. But it never leaves you.'

The physical commemorations of the sacrifices made by oil workers and their families to maintain the supplies of oil and gas from the North Sea are to be found in Aberdeen's Hazlehead Park, in the form of the starkly simple sculpture which Ann Gillanders referred to, and in the oil chapel in the Town Kirk, the East Church of St Nicholas.

Piper Alpha is always regarded as the sobering wake-up call for the industry, but Mike Jennings, for one, is not convinced that, in Kevin Topham's words, 'lessons had been learned'. He said, 'People talked about the Piper afterwards and how bad a platform it was. I didn't find it so, but there again I was new to the industry and I hadn't seen anything prior to Piper – just drilling rigs and I thought it was quite nice compared to drilling rigs. But I did look around the Claymore and think, "God, this is another accident waiting to happen." Corrosion – far worse than Piper to my way of thinking, and there seemed to be quite a slack attitude towards things, even after Piper. It couldn't happen again. That is what I said at the inquiry afterwards. There are so many people going around saying "This can't happen again," but listening to what people are talking about and what they are doing and going to do, mark my words, it will happen again.'

Human Interest or Business Interests?

[10]

After the Piper Alpha disaster in 1988, it took five years for new legislation to come fully into force. The new offshore safety division began operating in April 1991 and its responsibility for enforcing the regulations was enshrined in the 1992 Offshore Safety Act. Through the industry working party, Cost Reduction in a New Era (CRINE), UKOOA described safety management as 'an integral part of the reorganisation of the North Sea industry'.

All the companies' safety cases were finally submitted for 216 fixed and mobile installations by the 1993 November deadline. But the assessment and acceptance was an equally slow process. Some critics, such as Woolfson, Foster and Beck in *Paying the Piper*, maintained there was 'oil industry resistance' to the new offshore regulatory framework, finding it 'too prescriptive' and resulting in what they called, 'a policy of containment'. This had led to 'a gradual erosion of regulatory reform'. UKOOA later rejected the Woolfson findings and insisted there had been an improvement in accident statistics – 48 per cent down since Piper Alpha. But in 1996, the number of fatalities and serious injuries were again on the increase.

Immediately after the disaster, many of the companies were galvanised into modifying the elements identified as being at the heart of the systemic failures. Alan Higgins said, 'Within six months of Piper Alpha [even before the Cullen Inquiry], Chevron spent £1 million installing subsea valves on all the main pipelines on the seabed. They calculated the inquiry would legislate for this, but it never happened.' Occidental had also acted after the disaster, according to Jane Stirling, and Piper's feeder platforms were shutdown during the investigations. 'The decision was takento install emergency shut down valves subsea so that feeding a fire could never happen again. Eventually, there was also a separate linked accommodation platform.' The Piper successor, Bravo, was built, with the accommodation module out of harm's way, at Ardersier and installed in 1991. Mike Jennings saw the changes on Claymore. 'There is a proper escape route built from the drill floor to a safe area, so you don't have to cross the deck. But for months I used to wander

round checking all possible escape routes. I was never really happy and first chance I had I moved to Piper Bravo.' It was claimed by UKOOA that by the time the Cullen report was published the companies had spent about £1 billion constructing a new safety regime. A further £5 billion was invested in the following ten years.

Naturally, across the oilfields the shock waves had spread and many oilmen went back to the beach never to return, while others thought long and hard about continuing. Contract painter Graeme Paterson believes that if he had really thought about it he would never have gone offshore again. 'The Piper showed how bad things were, but any platform you go on to you are still sitting in a Piper situation. If something goes wrong, the stuff you are dealing with, gas and oil, is a hazard. After Piper, I seriously thought about giving up, and then I thought, "What am I going to do on the beach?"' Graeme had every reason to be concerned. Despite the 'wake-up call' of Piper Alpha, the offshore tragedies continued. The year 1990 was bad – in June, the *Ocean Odyssey* rig blew up and sank, taking with it the radio operator, who had stayed at his post to the last. A month later, six men died when their helicopter hit the Brent Spar storage unit. Over the following four years there was a combined total of more than 400 fatalities and serious injuries offshore.

Nevertheless, according to those responsible for overseeing the new rules and guidelines, there was an unmistakeable evolution in attitudes to safety and to regulated safety management systems among employers and employees alike. Sir Ian Wood of the John Wood Group, accepted neither the 'accident waiting to happen' argument about Piper Alpha, nor the criticism of the cost cutting after the oil downturns, or of the safety regimes. 'My people very closely investigated the extremely unfortunate set of circumstances and the negative coincidences. I think the inevitable conclusion was that certainly some operators – and I don't think it was anything to do with money – simply weren't being vigilant enough, didn't focus enough on the key systems and didn't spend the time and attention on the way they worked. Since then, the safety regime has been chalk and cheese. There is no other industry in the world where there is so much attention paid to safety. I take a personal interest in it on every contract. There is now a huge amount of time and attention paid to safety.'

Mike Waller, who rose from roustabout to operations manager at Shell, believes that in his thirty-three years in the industry the new attitudes to safety constituted the biggest change. 'I am not sure the public appreciate how much serious effort the companies put into safety. Certainly it is always challenged when something happens, but, genuinely, people are trying to do something about it.' Mike worked for a time in Norway. 'We shouldn't think we are any less well-prepared safety-wise here than on the Norwegian side. The regulations are certainly there, but in reality in everyday life, they are not any better than we are.'

Ray Craig, who is involved in training OIMs, has a concern about the nature of the post-Piper approach to safety. 'I think that overzealous is the wrong word to use, but there is a generally felt view – and I can't disagree with it – that there is an overcautious approach, which removes initiative. The trouble is you can say that and then see a guy dead or injured. It is certainly a totally different environment now – substantially better – and I think it is wrong to go into any sort of criticism of overkill in it.'

The roles of the safety representative and the safety committee which Lord Cullen had urged the companies to support and encourage have always been grey areas on the installations. For Jake Molloy, it took a frightening health and safety incident on a Brent platform the year after Piper Alpha to turn him into an active if unofficial safety rep. 'There was this unit we used to top up with oil, and if you got a build-up in pressure you knew it had to be shut down, as well as the whole platform. Just as the guys were queuing up for their New Year's lunch, the unit blew – like a coke can – twisted in half with metal flying everywhere. We were at muster for over six and a half hours. The management blamed the OIM and his deputy for being drunk. At Christmas offshore you got a little bottle of wine and two cans of beer which you had to sign for, while management looked after themselves. I don't believe the OIM was drunk. He handled the situation very well, everybody was briefed with constant Tannoys. Unusually, it was a very foggy day, but he had decided they would try to get people off by helicopter. He planned to shut the platform down and call in the *Stadive* and the standby vessel. Then he monitored the situation, waited till the area was cleared and put the platform back to normal. That OIM and a few of his colleagues were moved and then subsequently dismissed. There was no industrial tribunal. But that was a failure by the management because that vessel should have been shutdown for repair. Incidentally, that actually led to the banning of all drink offshore for the entire North Sea.

'That was when I realised it was just a sham, all this management speak about health and safety. There were no safety reps, but I decided to fulfil that role for the deck crews, painters and carpenters. I was fortunate in that Dick Murray was supervisor – if it had been any other supervisor I would have been run off. I did it with a high profile, pulling guys in as witnesses. This was just before the regulations in 1989. The OIM said. "You clearly want to get involved, so how do you fancy standing for election?" I got all the training they provided, a week off to do a five-day course and I was the first safety rep, elected in February 1990. At that time organisation offshore was just starting. OILC was in its infancy. No one else was doing it – but they did quite rapidly after the sit-outs.'

With the transfer of power to the HSE, the number of inspections offshore increased. The officials were not always welcomed by the experienced platform management. John Nielsen explained that OIMs were to all intents and purposes 'employed' by the HSE. 'They have to give permission for you to become installations manager. So you are the one they go for; you are 100 per cent responsible for the safety and well being of everyone on board. Chevron were well ahead on the safety front, covering a helluva lot of stuff long before Cullen. Safety is basically common sense.' Another Chevron manager, Alex Riddell, felt they were caught in the centre. 'Some of the older OIMs found pressure was being put on them from the town because the operators had suffered a big culture shock. Even in Chevron, with its American management, we were always nervous when the HSE inspectors came out. The OIM is the meat in the sandwich. The HSE is on one side trying to pressurise you and then you have your management saying, "Don't rock the boat." So either you are your own person or you submit and your management will beat you on your head even worse.'

The new safety climate has also spawned more administrative work. Roy Wilson worked latterly for what became Exxon Mobil as drilling engineer manager in Aberdeen, and he has seen tremendous changes since he first boarded a drilling rig in 1973. 'When you see now what people have to do to keep those places safe, it is extraordinary. You can hardly do anything without doing an accident assessment. The annual number of accidents and incidents has dropped tremendously. It is a very highly regulated and safe industry now. What doesn't quite jibe with that, however, is that oil companies are squeezing people to cut costs. Bureaucracy has gone up and people are loaded with regulatory paperwork. I had to call my big boss in the US on a Sunday night if someone cut a finger. They were paranoid about safety because a lot of their bonuses are tied to safety figures. I used to call it "feeding the monster" – they are so overloaded with paper I don't think they are capable now of making these places any safer.'

Shell's Mike Marray had spent most of his career working overseas. When he came back to work in safety in the 1990s, he found the general public still sceptical about the industry's commitment to safety. 'Safety is paramount now, which I have always found difficult to explain because many people seem to think the industry doesn't care about the environment and isn't particularly mindful about safety. Nothing could be further from the

truth. The drive from the most senior management in Shell was very much safety, environment, reputation – but the one thing you talk about offshore is safety. It is only in the last year or so that when companies said the number-one priority was safety people actually believed they meant it. That has resulted in a big improvement in safety performance.'

Australian Ralph Stokes, who was employed by the Wood Group offshore, believes it is now a little bit safer than it used to be. 'But it is not as safe as it could be. Sometimes oil companies tend to pay lip service to the safety side. If it is convenient, then they will go all out for safety, but if it is not, then it goes by the board. Having said that, everybody that I worked with was fairly safety-conscious and it is good to have safety officers roaming around all over the place.' Even the relatively new young Shell engineer Sarah Wingrove sees a change in her time offshore. 'Safety is definitely improving. When I was on deck, the management did talk the talk, but they really didn't do it very well. They told you to wear helmets and glasses, but they took them off themselves. Then when things went pear-shaped, they would be shouting and bawling. There is very much less of that in the last few years. That has all been tightened up.'

Leaving aside the design, structural and mechanical failures that precipitated Piper's destruction, two other sets of circumstances stand out from the Cullen Report: the failure of the management on Claymore and Tartan to turn off the flow of oil and gas condensate and the total ineffectiveness of evacuation plans. The latter was what concerned Kevin Topham when he asked why no lessons had been learned from the first industry disaster.

What wasn't in the emergency plan, as Mel Keenan points out, was: how could people save themselves? He had returned to the North Sea to work with Elf as an offshore training co-ordinator. 'One of my first jobs was to organise training for fire teams on Piper Bravo in 1993. My own RGIT certificate was due for renewal, and as part of the course I went out in the bay in a lifeboat. I asked a guy beside me what he thought of the training. He said, "It is a waste of f—ing time. I'm out here in this lifeboat and they are going to show me how to be a coxswain. I am never going to see the lifeboat, and why have forty-five peoples' lives in my hands?" He said that if no coxswain appeared he would probably try somewhere else. So I began to wonder if the training was targeted where it was needed, in those days a heretical question. But I went to UKOOA meetings and asked, "Have we got this right? We are training these guys to be firefighters, but are we training them how to get off?"' Mel was appointed by UKOOA in 1997 as project manager to review their guidelines for competence in training for emergency response under instruction and for evacuation. 'The guidelines are backed by

Plate 99.
On Shell's Auger TLP, a fire team in training. Pete Blasinghame holds the nozzle while Bo Evans and Richard Cox control the hose under the supervision of Kelly Bergeron.
(Shell International Ltd)

trading standards for every single emergency role – everybody is thoroughly trained. That was a sea change and it became a kind of mission for me.' Mel is now chairman of RGIT Montrose.

Another striking feature of the disaster was the fact that Piper's sister installations kept pumping fresh fuel into the furnace the platform had become. There was no total shutdown of the pipeline until almost two hours after the first explosion. The inquiry was told neither of the installation managers on Claymore and Tartan was prepared instantly to close down production. A commonly held view among oilmen was expressed by Graeme Paterson: 'It was the very fact they never shut off the valve and the stuff was still pouring out. And of course, the Piper was a giant pumping station. What sticks in a lot of guys' throats is that these platforms wouldn't shut down until they got permission from the beach. You are speaking about lives here – stop the bloody things pumping – but they wouldn't do it.' This is the issue unions call 'the culture of fear' which lies at the very heart of the offshore safety regime: the unwillingness to shut down without orders from onshore for fear of being sacked or at the very least run off the installation. Ronnie McDonald, who came from the Merchant Navy, started work as a crane operator in the Highland fabrication yards in 1975. He went offshore as a rigger in 1980 with a number of companies, including the Wood Group. He was a trade union member. He recognised that despite the changes, the men in the oilfields were still vulnerable. 'Some aspects of employment law did apply. People dismissed by a shore-based company could take them to a tribunal, but the law in the North Sea was the "NRB" law – "not required back". A contractor employed you for a trip, but the operator would notify your employer if you were required back. If you were mouthy, even with a legitimate complaint such as safety, it made no difference. If the invitation was withdrawn, no explanation was required. The operator had complete influence. You could take your employer, the contractor to a tribunal for unfair dismissal – but it would be sufficient defence for him to say, "I am following instructions from my client. I have no alternative employment to offer, so therefore in all the circumstances the dismissal is not unfair." And that was it.' Nine years after the disaster, a Mori poll for Shell Expro showed that 58 per cent of their employees believed speaking up could damage their job prospects. In a comment on the poll, university researchers claimed a fearful 'mobile workforce with very few employment rights meant full implementation of the Cullen Report was not possible. The dogma of speak up and you are out – the NRB system – remains in the industry.'

Piper Alpha, in the opinion of Alex Riddell, was an outcome of just such a work culture in the operating company. 'In a video they made, an OIM on one of the other platforms said he wouldn't shut down until he got someone on the beach to tell him. Now that is a cultural issue and it still exists because the OIM offshore hasn't got the balls to do it. He has the power – but is he prepared to challenge his boss? You just have to say the decision you made in the height of the storm was your best decision. You can always turn on the tap again. I don't want to be sitting in a boat in the middle of the North Sea and be asked, "Why didn't you shut down." So always err on the side of safety and shut down.' The UKOOA guidelines produced in the 1990s gave the offshore installation manager the authority to act autonomously. Mel Keenan, who had a hand in drawing them up, said, 'They used to say at the BP refinery at Grangemouth, "If you are in doubt, shut it down – if you are wrong, we still have a plant to start up in the morning. Try to keep it going and the plant could be in smithereens with dead bodies all around."'

Nevertheless, there is still a belief prevalent in the industry that the pressures from above – overt or covert – to keep production running have not eased. Jake Molloy is sympathetic with the men in the middle. 'I think they have the most difficult task. The commitment at the top is unquestionable. Lord Brown of BP and Michael Brindle of Shell don't make the

statements on health and safety and not mean what they say. The guys at the bottom are likewise committed, but maybe lack the education to push their case. But the men in the middle are pulled between the rock of finance and the hard place of health and safety – even more so now. Offshore, instead of the engineers, the accountants run the show.'

The Amicus union organiser Graeme Trans regularly comes across NRB cases. 'It is sad to say that a number of membership forms have a note on them, "Do not tell my employer I am in the union." There is a fear of blacklisting which I believe still goes on. There is still the NRB issue. It is morally wrong. The contractor will do it because if they don't it it can be financially detrimental to renewal of contracts. And if it comes to a tribunal the employer, the contractor, will say we don't have a problem, it's the operator. But you can't take a third party to a tribunal. One tribunal chairman said publically, "This is morally wrong. The only way to change it is to change the law." And that is what the unions will have to work at.'

This question has been the subject of discussions at UKOOA, according to director of communications, Steve Harris: 'There is a new code of practice which tries to get some clarity and transparency around this issue. It is obviously something that has niggled away. No one would say that if you have somebody who is a real problem you have to keep him on your platform. But the contractor is ultimately responsible, and this needs to be looked at. I think under the new system people will be seen to be behaving legitimately or not. Then the people concerned will have a case.' Steve says that of all the issues in the industry, NRB is fairly low grade. 'But it does lead on to the bigger issue of safety. Do people feel they can hold their hands up and say, "Stop?" Sitting in safety meetings five years ago there clearly were people who felt you couldn't do that, but a helluva lot of work has since gone into trying to change that culture.'

The Offshore Contractors' Union also maintain they have tried to dispel what chief executive Bill Murray describes as the NRB myth. 'This may have happened in the past, but it is certainly not anything we would condone now. We have an open offer to the unions: if they can identify any instance of it, we will take it up, because it hampers what we are trying to do on the safety front and the general relationships. We want the workforce to tell us. Five or six years ago, every managing director of every OCA company wrote to his workforce, "Look, if things aren't right, stop the job. Anybody stopping the job on safety grounds will get our full and unequivocal backing." We have tried everything we can, but we still can't get past the bogeyman under the bed even though everybody knows he is not really there.' Within contracts there is the right to ask people to be withdrawn from a platform. 'It is how that right is exercised that is crucial, and we have been working with the unions to reach an acceptable procedure to investigate these allegations and produce a remedy. So I really don't see that the situation hasn't improved.'

Related to the fear of the consequences of stopping production are the implications of the accident-free bonus, referred to by Kevin Topham and Roy Wilson. Graeme Tran cited an example. 'One of our members was injured offshore when some equipment fell on his head. They wouldn't remove him from the rig but gave him painkilling injections. The OIM saw him later and the guy said he hadn't been able to get back to work. "I am not right, I have to get off." The OIM said to the medic, "We will have to discuss that." The medic said in front of everybody, "You know what you have just done? I have to report that and that now takes us over and we have lost our bonus." Safety bonuses have to be done away with. Their time is spent.' Such incentives concerned Piper survivor, Mike Jennings. 'A lot of measures were a sham. If you saw a guy doing something wrong, you had to tell him and then fill in a form; so many forms per shift per trip meant points towards your annual bonus. This made people tell tales to get money and some filled in forms just for the bonus.' Mike, who ended up on Piper Bravo with two jobs – helicopter information officer and radio operator – retired after suffering a slight stroke. 'My wife thinks the pressure contributed to my stroke.'

The post-Piper concentration on a new safety culture was somewhat overshadowed in 1993 by a totally cost-led initiative which emerged from a claim that the extraction of oil from the North Sea was four to six times more expensive than anywhere else. The companies said they needed cuts of up to 30 per cent over three years. Sponsored by UKOOA, this was Cost Reduction in a New Era (CRINE), but it did not meet with universal approval. Jim Milne, boss of service and engineering company, the Balmoral Group, said in 1999, 'I believe the industry may regret adopting such a vigorous attitude to the continual cost-cutting following the success of the first CRINE initiative. The first 20 per cent in cuts was entirely acceptable . . . However the subsequent drive for a further 20 per cent cut and yet another round of cost-cutting was just too much . . . Many SMEs have been forced to the wall while others have had to compromise or consolidate.'

Mr Milne's warning came at the height of a fresh global oil crisis in 1998. The price of a barrel fell to $12.5 that summer, plummeting to an all-time low of $9.14 in December – worse even than in 1986. Once again, the UKCS vulnerability to OPEC, which had allowed oil stocks to mount, was seriously exposed, and non-OPEC governments, the UK, US and Norway, were apparently unable to act. Exploration was badly affected and new

exploration and appraisal wells were down by half. As crude prices continued to fall, UK oil revenues hit their lowest level for fifteen years, with a 37.8 per cent drop on income from sales. New projects approved in 1998 were worth only £1.5 billion – barely a third of the proposed £4.2 billion investment. While oil revenues were falling to just £19.1 million, production – as in 1986 – had risen. In comparison, however, the impact on employment in Scotland was less traumatic; of the 15,000 workers made redundant, 1,500 to 2,000 were based in Aberdeen.

Relief came in March 1999, when OPEC finally cut quotas and prices rose to between $15 and $16 a barrel. Oil revenues responded, but drilling remained seriously depressed. Despite the crisis, a record sixty oil-only fields were in production in 1999. Then, just as in 1986, the major oil companies stimulated recovery; BP announced expenditure of £600 million, while Shell boosted investment by 50 per cent to £800 million to cover six new drilling developments. The recession was over and the industry learned the hard way to base future planning on lower oil prices.

CRINE achieved its target and was replaced by a new joint Government and industry body, LOGIC (Leading Oil and Gas Industry Competitiveness), designed to create more efficient offshore supply chains. The Government oil task force was replaced by PILOT, a more potent joint group.

In an ageing province like the North Sea, oilmen now question the safety of the original platforms. Some have already gone. From 1988 to 1998, eighteen installations were decommissioned. The most controversial was Shell's disastrous attempt in 1995 to dump their Brent Spar concrete storage unit in the Atlantic, miscalculating the public relations skills of the militant environmental group, Greenpeace, who fought a masterly media campaign forcing Shell to drop their plan. The activists' assertions about toxic residue inside the buoy were later proved unfounded, but it was too late. The unit was redesigned as a pier outside Stavanger. An international abandonment policy, agreed by the Oslo-Paris Convention (OSPAR) in 1998, now prohibits dumping and requires installations weighing less than 10,000 tonnes to be returned to land. The Spar debacle also refocused attention on the considerable impact the industry has on the natural marine world around it. Inevitably there have always been environmental concerns about the disposal of chemical waste, the handling of drill cuttings, and the possibility of oil spills and pollution. Such issues are now heavily regulated and the industry has to pursue rigorous management strategies to alleviate any damage. Environmental case studies are also an integral part of the development acceptance of new fields. By most accounts, the industry appears to be maintaining a satisfactory record, complying with European directives on

Plate 100.
July 1997 – the storage unit, Brent Spar, is dismantled in a Norwegian fjord before being incorporated as part of a pier at Stavanger. *(Shell International Ltd)*

good environmental housekeeping.

The other environmental issue was the interaction with their North Sea neighbours, the fishermen. After the Seventh Licensing Round in 1981, a code of conduct was devised to cover the most sensitive area, the inner Moray Firth. But what aggravated the fishermen was the prevalence of oil-related sea bed debris, so UKOOA eventually financed a compensation fund. But the fishing industry was also concerned about the hazards of permanent obstructions. The most tragic case was the loss of the Arbroath vessel, *Westhaven*, in 1997, when she snagged her trawl board on an oil pipeline. The four-man crew died when the vessel was dragged under. Since then, the

THE OILMEN

Plate 101.
Brent Charlie – the platform and the other Brent field installations underwent major improvements in a £1.3 billion five-year re-development programme.
(Shell International Ltd)

fishermen and UKOOA consult electronic charts which warn of the location of subsea hazards.

In contrast to the abandonment process, other fields are being restored to life. One field, which originally defined the potential of the UKCS, has been refurbished in the largest single engineering project in the North Sea's history. Shell's Brent was redeveloped in a five-year programme, which began in 1993 at a cost of £1.3 billion and involved some 3,000 people, extending the declining asset's productive life by another ten years. Still embedded in the seafloor, however, are other first-generation giants, battered and much depleted. Their alleged neglect is the source of another safety fear offshore. Former general manager of Chevron UK, Alan Higgins, is chairman of the Institute of Petroleum. 'Ninian must be in a hell of a mess now because it is being maintained with eighty or ninety people. We had 400 and we couldn't keep up. That brings you right back to another disaster just waiting to happen. I worry about that, having lived through Piper Alpha, and it must be true of a few of them. They all have new or different owners, all making money and doing it by cutting back on the workforce. Maintenance must go by the board and they will just run the thing into the ground. People get complacent, but these things go in cycles and I'm not so sure it's not coming round again.'

When Graeme Paterson started going offshore to platforms like Forties Charlie, he was one of thirty painters on day and night shift. 'Now there are maybe two, still day and night shift. Obviously you can't do the same amount of work, but offshore stuff just doesn't last and it must be treated on a regular basis. Nowadays, they seem to forget that. So the fabric and the maintenance side, a health and safety issue, is down to just replacing parts. It's all about cost. These companies can run these fields because they can cut to the bone.' Two companies whose elderly installations were criticised by both Amicus and the OILC, Shell Expro, and CNR who took over Ninian South, deny there are problems. But Graem Tran still believes the maintenance question is one of the biggest facing the industry. 'They didn't realise the backlog out there.'

UKOOA's communications director Steve Harris says the companies regard maintenance as a big challenge. 'Most of the older ones were never designed to still be there, so there is obviously an engineering challenge to keep them going . Everybody accepts that in the next couple of years asset integrity, as it is called, is going to be a crucial issue. I know the HSE think it is. Certainly it is at the top for Step Change and UKOOA.' Bill Murray says maintenance is the OCA members' responsibility. 'Operators rely more and more on contractors for engineering and maintenance scheduling.' He also sees it as a challenge. 'As the things get older they need more maintenance, but we also understand them better. We know what is likely to go wrong and where interventions are needed.'

The onus falls ultimately on the small companies, as Jake Molloy discovered. 'If you ask them off the record how they are getting on, the reaction would make the hair on the back of your neck stand up. They are being squeezed so much and forced to take so many risks that, for many, continuing to operate in the North Sea is no longer viable.' Bill Murray did not accept that the economic crises had affected attitudes to safety. 'I would say genuinely I have not seen any reduction in standards driven by cost. In fact, if you look at what was happening in 1985–86, when the industry was at its lowest ebb because of the price shock, companies were still investing in safety. Even today the investment in safety is holding up.'

In the years since Cullen the combined rate of incidents involving fatalities and serious injuries declined – in 1993 the figure was almost half that of the previous year – but the number of deaths in 1992 was magnified by one single disaster, the ditching of a Bristow Super Puma near Shell's

THE OILMEN

Cormorant Alpha platform during a shuttle trip from an accommodation barge. Eleven of the seventeen men on board lost their lives. Experts later claimed that because of the adverse conditions the routine flight should never have taken place. It had been 'a commercial decision'.

A common answer about the importance of safety is that apart from the unacceptable human cost, accidents and fatalities are bad for business. The challenge is to achieve a balance between business and human interests. In 1997 the industry launched the Step Change programme to try to cut accidents by half. A series of measures were introduced: a safety representatives' network, safety leadership training, a common induction process and an offshore passport issued in 2000. It is a cross-industry initiative, involving UKOOA and the OCA. Graem Tran sits on the Step Change forum and on the Government task force. He is aware other unions don't support the safety initiative because legislation bars trade union safety reps offshore. 'The ideas are getting better. The vantage card, for example, a kind of offshore passport which records how much time you have spent offshore. But they seem to be losing out on the basics. Somewhere down the line there will be a major fallout.' Figures released three years after the launch by the HSE showed there had only been a 24 per cent improvement overall. Later figures revealed major injuries increased by more than a third, with three fatalities in 2004. Offshore unions say real figures are higher because many injuries go unreported. The Amicus organiser claims 'The HSE have a lot to answer for'. He said, 'They are supposed to spend a lot of time and resources on investigating, but they are toothless and not prepared to challenge the companies. I believe the offshore division in Aberdeen has a good team, but let them do the job they are supposed to do.'

The Step Change chairman, at the time of writing, is Alison Golligher, UK managing director of Schlumberger Oil Services. She began working life on a land drill in the jungles of Indonesia, where safety was not afforded the same importance. In the UKCS, she said, there is a huge difference. 'I feel the oil and gas industry takes its safety very, very seriously – as it rightly should. Can we do better? Yes, as long as there is still one person hurt we can always do better. Is the industry trying? Yes, I think it is trying hard and it is a constant grapple between the practicalities of running a business, protecting your people and obeying legislation. The most important duty of Step Change is helping people to choose what is most important and what is going to have the biggest impact for those who are at risk. I have no doubt all of my colleagues – the managing directors – are completely committed. The worst thing is to wake up in the middle of the night and get a call. I think we all have work to do within our companies; it is the ongoing challenge to make sure that commitment is echoed throughout every level of management.'

Alison is very clear on the controversial issue of the man on the spot offshore calling a halt if there is any hint of a risk to safety. 'Safety goes out the window in some circumstances and that kind of behaviour is not acceptable. I think some of it is self-imposed and some is caused by a lack of clear, supported and demonstrable actions. I am glad to say my guys will say if they want to stop the job. If somebody offshore tells them they are not keen for them to do that, they will call in and I will call offshore and say we are not going to do this because it isn't safe – and they have done that every time I am aware of it. The customer has always been very supportive.'

The Schlumberger boss is very concerned about the issue of communication downwards. 'There is a very big disconnection in communications through the ranks, to the extent that the industry should be asking itself some hard questions. Statistics have improved dramatically, but are they where we want them to be? No. Would we like to have no fatalities from now to the end of the North Sea? Absolutely. Nothing we should ever do should take away from that. Do we accept them when they happen? Absolutely not. Are we always as effective as we would like to be in making sure that the preventive actions really work? I would say, no, we are not, and that is where we need to spend a bit more time and effort. I think the industry tries hard, but it can be better and more effective. There is never an end to it and there never will be – human nature just isn't like that.'

The demographic structure of the offshore workforce is beginning to concern many people. The average age is now fifty, but there are older men still working. Derek Stewart, a ROV contractor, for example, is one of the oldest at sixty-seven. So the industry is constantly trying to attract younger people. One immediate problem is that these young entrants may be unaware of what Piper Alpha meant for offshore safety. Some companies, such as Total, therefore show a video of the Piper Alpha Tragedy during training courses. Jane Stirling, who was called in on the night of the tragedy, said some of the audience are visibly involved when they see what happened. 'I can remember as a little girl how the Aberfan disaster affected me, and that's how these young people see Piper Alpha. The difference is they are working in the industry and they need to know. The feeling is, okay, we haven't forgotten, but we have to remember the lessons learned from it. We don't want these young people to be allowed to repeat the same mistakes.'

[11] Not Required Back

'OILC emerged out of Piper Alpha in 1994 – simple as that. I came home the day it happened and two days later my wife said, "There is a meeting about Piper in the Trades Council." I saw Ronnie McDonald and spoke at the meeting. We started getting organised then. We saw that nothing was being done and these guys were being killed. So Ronnie and I started the OILC. It was for oil workers and run by oil workers and every one of us was long-serving offshore. People like that were involved and they knew what was going on. OILC has made a difference, because it has given the oil companies somebody to watch.'

Dave Robertson, a boilermaker from Motherwell, is a union activist from a trade that has a long tradition in organised labour, centred for the most part in Scotland's shipyards. He made no secret about his beliefs when he first went offshore in 1978, and when he experienced for himself the conditions men were working in – the dangers and the pressures – he began to try to organise the workforce on the various rigs and platforms he worked on. Since then he has been sacked, run off and blacklisted, although on the first occasion he was 'blacked' he was able to return offshore with another company. He now works onshore, but he is still a member of the OILC. Although many hundreds of oilmen also suffered under the draconian 'not required back' (NRB) system, Dave is not at all typical.

Trade union recognition offshore has never been an achievable option over the bulk of the three decades of the UK oil and gas industry. While industrial wars raged onshore in almost every sector of British industry, from the 1950s until Margaret Thatcher's Government changed the employment laws, the history of industrial relations offshore has been less clearly defined, largely because representation has either been sketchy or non-existent. There were sporadic disputes, but not until the birth of the Offshore Industry Liaison Committee and their 'summers of discontent' in 1989 and 1990 was there any co-ordinated organised industrial action.

When the North Sea offshore oil and gas industry was young and raw in the 1970s, there was no ambiguity about the issue of organised labour which so exercised the politicians and trade unionists onshore: there was to be none. The reasons at the outset were cultural. The oil fields were the closest Britain had come to reverting to a free market capitalist society since the heyday of the paternalistic ironmasters and mine owners in the nineteenth century, and the American rig bosses – with the attitudes they brought across the Atlantic – could have been their direct descendants. Jack Marshall, Conoco's vice-president in charge of international production in London during the Murchison and Hutton developments, best summed up their business philosophy: 'We tried to run our business the way we did in America so that there was no advantage to a man in being in a union. As this became apparent to the unions I dealt with here, there were fewer and fewer of them. We would actually give raises and we would go to the unions and say, "We have given these other people raises. If you would like to have them, do you want them included in your contracts?" If you run your business – and labour is such an important factor – it is an absolute necessity that the better it is, the better you are and the better they are. We worked on that philosophy.'

The British companies saw in the North Sea a unique opportunity to make a fresh start without unions in a labour relations climate which appeared to have had gone so sour in the onshore industries. Anomalously, the majority of the workforce appeared prepared to accept this. From the point of view of the workforce, Ronnie McDonald said that many of the men had brought with them bad memories. 'These guys came from areas where there wasn't any work, where traditional industries were shutting down – Tyneside, Teesside and the Clyde, where the shipyards were going down like nine-pins. So it was a tremendously valuable human resource and these guys were glad of the opportunity of further work. The oil companies milked this for all their worth.'

There were other workers, from the disciplines of the mercantile marine or the armed forces, who were new to industry and saw no need for unions. John Williams, a production supervisor on BP Forties Alpha, was asked by Jimmy Young, when the radio presenter broadcast his programme live from the platform in 1977, why there was no trade union representation at that time. John said, 'It is fairly new – I think that is one basic reason. A lot of the men, I think, are individuals, and that doesn't go too well with trade union representation.'

The international oil operator Chevron was another American company that would never countenance unions. Chevron employed former Naval officers as OIMs. One was Alex Riddell, who worked on Ninian. 'The union officials had access to the workforce, but no one ever signed up. That was

THE OILMEN

Plate 102.
Ronnie McDonald addresses a rally in the thick of the campaign for recognition for the OILC – the radical group he had helped to found as the first secretary. *(OILC)*

due to the Chevron culture and their "open door policy". They had quite a good human resources staff, who were very supportive and knew how to handle workforce grievances. But there were very few of those. The company followed the letter of the law and gave them a fair crack of the whip.' Alan Higgins, another ex-naval officer and Chevron OIM described the situation during platform construction. 'These guys were all 'here today, gone tomorrow' construction types: steel erectors, scaffolders, welders, boilermakers, you name it, all the black trades, and they would down tools at the drop of a hat. They all seemed to be ten feet tall, seven feet wide and built like brick shithouses. There was a storm one time and three great bozos came into my office and said, "That's it" and downed tools. "Get us ashore." And I said, "Wait a minute," but there was no reasoning with them and the whole lot went ashore. It took several days to resolve the dispute and even now I don't know what it was about.'

Charlie Brown, an old-style oilman who began in Shell tankers graduating to installation manager on Auk and Dunlin, says: 'I always told the men, "The reason unions want to get involved in the North Sea isn't to benefit you guys offshore – it is only to use you as a fulcrum to lever up the guys onshore." A lot of them had been members of trade unions onshore, particularly the electricians from the mines. These guys would tell you quietly they were glad to be away from the unions, which dominated the mines. My philosophy has always been: providing you give people a fair day's salary, look after them, don't bully them, don't threaten their jobs, then they are not going to bother with trade unions. A man joins a trade union for protection, trying to improve his lot, or if he is scared.'

Former Forties OIM Jim Souter said the men who came from the oil refineries were very unionised. 'They were inclined to go and see any union people who came on board. They felt more comfortable speaking to a union rep than to a supervisor or their manager. The management was the enemy really – that was an attitude they had brought from onshore. In fact, it was probably the OIM who needed the protection of the union.'

The offshore ambivalence towards unions is also illustrated by the fact that oilmen, generally, are regarded as notorious for ducking confrontations. Jake Molloy used to meet with platform management as the men's safety representative in the 1990s. 'I always made sure there were sufficient witnesses around to support me, because the one person you had to worry about most when the shit hit the fan was the guy sitting next to you. They are not renowned for their collective might, although they showed it in 1989 when they were given reasonable leadership guidance.'

So the fundamental issue is: are unions really needed offshore? It is clear that up to the 1990s, the offshore battalions were totally at the mercy of the operators and contractors. From the early days, a doughty champion of the oilmen was the Aberdeen-based North Sea Oil Action Group, led by an active trade unionist, Councillor Bob Middleton, later convener of Grampian, the 'oil region' council. In one speech in 1974, Bob called for the unionisation of the workforce. He pointed out that the oilfields were outwith Government industrial legislation, which only extended to the twelve-mile territorial limit. 'Two oil companies, Shell and BP, have said they are not opposed to unionisation, though, quite naturally, they will do nothing to assist, [but] it is the bulk of the charter rig operators, mostly American, who are actively opposed to trade unionism.' He said, 'These companies offer contracts of employment which, I am sure, would have pricked even the consciences of nineteenth-century coal owners.'

One early drilling crew learned that salutary lesson when they confronted American boss, Bob Rose, who was in charge of the first drillship to work in the North Sea, *Glomar IV*, and who ultimately became president of Diamond Offshore in Houston. He recalled later being faced by a demand for a 10 per cent pay rise by the rig crew. 'This totally incensed me. We were already paying good wages and there were plenty of people willing to go offshore. We told the captain offshore we would give the rise. Meanwhile, we hired a new crew, put them on a helicopter and flew them out to the ship. The captain fired everyone on board who threatened to strike. We had no more difficulties.'

THE OILMEN

The first full-blown strike in the North Sea, in 1975, was instrumental in bringing a change to the employment legislation. Dave Robertson was one of the prime movers in the dispute, which began on a SEDCO rig operated by Dixilyn for Conoco. There had been dissatisfaction over a changeover to equal time working which wasn't uniform. 'Then there was an accident and the OIM fired the assistant crane driver, who had been working with that crew for about eight years. We said to the boy, "Don't go, we are going to strike." The back-to-back crew also stopped work and they had to shut the rig down. I got fired for inciting and the OIM threatened to take us to court for mutiny. He said, "Come into the office in Dundee and get this straightened out." When we went ashore he fired us all. Well, we fought them for about twelve weeks and I got the Dundee dockers to stop Conoco's supply boats. Finally, I was reinstated and given my money back, but I was sent to Finland for a year.' When he came back, he found he was blacked.

Ronnie McDonald has another name for the system of blacklisting. 'I was told by a senior manager, "There is no black list – but there is a white list which you are all on."'

Black list or white, Dave Robertson simply circumvented the system and found another contractor who would send him offshore. 'After that strike, with the help of the Scottish Secretary Willie Ross and the TGWU, we got the law changed to include offshore workers in industrial tribunals. I was always very proud of that.' In 1975, under the new Labour Government, the anomalous situation ended when two new Acts of Parliament came in to force – the Employment Protection Act and the Health and Safety at Work Act. For the first time, offshore workers were covered. Despite this, the turnover of workers was excessively high (about 40 per cent), as thousands became disillusioned by the conditions and the wages. Fierce competition for labour in the developing oil fields then enforced a change and conditions began to improve. Seven days on and seven days off became the norm, with increased pay and bonuses.

But there was still no unionisation, and throughout the 1970s and 1980s a generally fruitless battle was waged by the eight unions concerned trying to get access to the oilfield workforce. The main thrust came in 1974 from a new Aberdeen-based body, the Inter Union Offshore Oil Committee (IUOOC), which replaced Bob Middleton's action group, and consisted of full-time officers from the pricipal unions. They had limited success. The body also carried a fatal weakness: the big unions had a history of internecine warfare over the demarcation of membership, so there was constant conflict. A further drawback was that the operators organisation, UKOOA, refused to recognise the IUOOC. The group, set up in 1973, maintained it was purely an advisory trade association, with no powers to negotiate on industrial issues concerning individual members or the contractors.

Then the new Energy Secretary, Tony Benn, took a hand and called a meeting in 1976 between the two bodies and from it a Memorandum of Understanding emerged, giving IUOOC unions transport and access to the offshore installations. But many of the operators continued to stall the unions and the number of union visits offshore until the 1980s was sparse, resulting in sporadic union coverage of only parts of the North Sea. A typical company response quoted in *Paying the Piper* came from Britoil to the IUOOC: 'We do not propose to have any discussions with trade unions on recognition for our offshore fields unless we receive a clear indication from a specific proportion of our staff that they wish to discuss some form of representation.' The TGWU's marine organiser, Mel Keenan, explained why the visits never really worked. 'The union officials would be there for a day or two days, but they weren't allowed to go and talk to people. They just had to sit in a room and wait for them to come. People were reluctant to do that.' He said that the unions also used to all go out at the same time. 'That put most of the guys in a quandary – which unions should they join? The understanding was that one union would represent all the interests involved. But it didn't work like that because the individual unions needed revenue from subscriptions.'

Mel had formed good relations with Occidental after he was allowed to recruit for members at the Flotta Terminal on Orkney. 'At that time, the oil industry paid more than any other industry in Scotland and Oxy paid 20 per cent ahead of that.' They were also more relaxed about access to their platforms. The T&G already had members on Piper, so they let me go to Claymore and I didn't have to go through the IUOOC.' The Piper Alpha staff ended up covered by agreement with two unions, the TGWU and MSF.

To go back to 1976, when the pressure towards first oil was building up and the demand for workers reached new heights, the construction and development phase was estimated to require a total of 12,000 highly skilled tradesmen. So, the oil companies felt vulnerable to industrial action and were open to a pact with the unions, the North Waters Offshore Construction Agreement – more commonly known as the 'Hook-up Agreement'. It sanctioned collective bargaining rights, the introduction of shop stewards and one offshore pay level subject to annual review. The IUOOC was not involved. The deal was initially successful, but once the major hook-ups were completed, some contractors began to make separate arrangements with non-union firms. The big test came in 1978, when a group of platform

THE OILMEN

Plate 103.
Jake Molloy, secretary of the OILC. *(OILC)*

engineering construction workers returned to the beach in a lengthy and bitter unofficial strike over the work pattern, pay and a restructuring of the hook-up. After nine weeks their union – concerned at the threat to the official agreement – persuaded them to go back to work. They had gained nothing and their failure demoralised the oilfield workforce.

'As soon as the first oil came the operators de-recognised the unions,' said Ronnie McDonald. 'That was because nine-tenths of the guys had been paid off. They maintained the core workforce on the platforms in the production phase so that the unions had no power. They divided the workforce into two – those who were well looked after and the hoi polloi. As was subsequently demonstrated come 1984 and 1985, the process was accelerated. Not only were your employment rights stripped from you, but with a succession of wage cuts there was nothing you could do about it.'

The IUOCC was also riven by factional battles. Elsewhere in the industry, the divers had formed their own union, while helicopter pilots in the main joined BALPA – the airline pilots' association.

In 1984, the urge to unionise was reawakened by a group of engineering construction workers who called themselves 'bears' and who produced a radical publication, *The Bear*. The group has been regarded as the prototype for the OILC. Eighteen months on, however, the industry imploded with the collapse of the oil price. By the time it hit the $10 mark, about 11,000 jobs had gone – and with them the latent dream of union rights. 'That was when you saw the smaller companies disappear,' said Jake Molloy, 'and the big boys coming in – Amec, Brown and Root and so on. I was made redundant. The company had just signed a two-year contract and they had decided my role could be better done by the catering company. Instead of £6 an hour I could get a job with the caterers at £3.25 or four quid an hour – putting me back six years – so I said no, and that was when I started getting a wee bit uppity.'

Mel Keenan was involved with COTA, the offshore catering traders' association. 'When the oil price collapsed, guys who had been working two on and two off had to work three on and one off for the same money. That led me into arguments with Oxy, because their partners insisted on a cheap catering bid, cutting the camp bosses' pay by about 40 per cent. Really quite savage, so I organised a protest. When the next bid came up, I phoned the company's HR guy. "There is somebody I want you to meet. He's the chairman of the dockers' branch in Aberdeen. If you take a cheap bid on your catering, you won't move as much as a nut or bolt through the port of Aberdeen." So no more cheap bids were accepted.' There was, however, a tragic sequel. 'Kelvin Catering got the next bid and the crew – many of them union members – were on only their second trip on Piper Alpha. Not one survived.' In 1974, following the Health and Safety Act, industrial safety representatives were appointed – but not offshore. Mel Keenan attended an IUOOC meeting when they discussed using safety as a Trojan horse, as a way in for the unions. 'I disagreed with that approach. Safety is too important to be subjugated to some other hidden agenda and I get angry when people mix the two up.'

As Dave Robertson said, the unofficial oilmen's 'union' was born out of anger about Piper Alpha. 'There was no mechanism for improvement offshore. In 1986, they started cutting the wage rates. In 1977, I was getting £5 for the first 40 hours and £7.50 every hour after that. From 1988–89, I was getting £6 for working 12 hours with another 50 pence on top. You couldn't refuse overtime. If you were working 12 hours you had to work 15 hours. If you didn't work it, you were off. You had no influence in your own safety. See if you were working in Richards, the carpet factory, under the law you had a trade union-trained safety rep. In the North Sea you didn't.'

The hook-up agreement was renewed in 1989, with the warning that there would be no new contract the following year. The offshore activists realised they would have to organise on their own behalf, and the OILC was born initially to educate offshore crews to the need for industrial action and to support the AEU and the GMBWU – at a time when the tough laws shackled Britain's unions. 'Ironically,' said Ronnie McDonald, who was the first secretary, 'the interest group most offended by our actions was the unions.' A complex four-cornered struggle intensified between the official unions, the OILC and the operators and contractors over union recognition

Plate 104.
The 'summer of discontent': the 'Bears' – the offshore construction workers and their supporters – on the march. *(OILC)*

and the right to collective bargaining. And so it has continued to the present time, with the OILC eventually achieving official trade union status in 1991 as well as recognition by the HSE, but not by the STUC. The OILC had begun by attacking the construction side where the derecognition agreement on first oil ashore had been dismissed. Then the impetus broadened to include other interests. 'The main unions were by and large supporters, but when we eventually got the offshore workforce together, the unions themselves couldn't hack it.'

In 1989, the OILC took their first action during the hook-up on Shell's Tern project, where Ronnie was working. The one-day token sit-in was timed to coincide with the opening of the Cullen Inquiry in Aberdeen. Later came the first of the so-called 'summers of discontent', which involved nineteen sit-ins on platforms over a nine-week period. The contractors and the companies were more or less conciliatory in a number of cases, promising to allow union recognition – if that was what the majority of their employees wanted. Some of the workforce gained improvements in wages and conditions. These strikes gave the men confidence in their ability to force change, and the following summer the OILC orchestrated another campaign specifically aimed at the FLAGs pipeline system in the East Shetland basin, beginning with an overtime ban and then a series of sit-ins on key platforms. The public were kept informed of the oil sit-ins through the OILC lively magazine *Blowout*, which continues to flourish as a radical oil industry commentary.

'The first day,' said Dave, 'we all got fired. Shell took up the running as the biggest employer, while Chevron, Mobil and the rest just got extra videos out for the guys on strike. So we had a 24-hour, 27-day sit-in on that Shell platform in the East Shetland basin. We had asked to be represented by a trade union of our choice and we wanted union-appointed safety reps and shop stewards, but they weren't going to have that. We were first item on Sky News. Then Saddam invaded Kuwait and the second night we were out of the news. If I had got a hold of that Saddam I would have strangled him myself. I got blacked when I came back. There were about three or four of us never got back.' Dave again managed to get offshore for some of the smaller contracting companies, but then came onshore permanently.

Shell took legal action and the OILC defended the action. The court ultimately ruled that the oil company were being disproportionately disadvantaged by the occupation, but refused to accept it had been illegal – ostensibly leaving room for further industrial action. The men came ashore – some of them still defiant. 'One thing we knew was going to happen was the effect on wages,' said Ronnie. 'Within eighteen months, all of the wage cuts of the 1980s had not only been restored to the previous levels but were also massively enhanced over a seventeen-month period. No contractor employee got less than 41 per cent – some actually doubled their money. After the first round of industrial action the minimum rate went up £7.50 an hour. Helicopter landing officers (HLOs), who had been paid as low as £4 an hour in April 1989 were on £8 an hour by September 1990. The reason was simply to buy the guys off. The more they perceived the development of collectivism, the more worried they got. By any measure of success there was certainly impact.' The OILC founder claimed that a year later, the most powerful of the unions cut a deal with the employers over a new construction programme and promised to 'wipe out the rank and file movement – to disown them'. The oil workers were actually barred from the annual TUC gathering in Brighton. At a highly charged meeting, the OILC decided to apply for certification as an official trade union, which came through the following year. 'By 1997, I had basically burned myself out and the OILC needed to be run by the guys offshore. So I didn't stand for re-election.' Jake Molloy became secretary. As well as being a safety rep on the Brent field, he had also acted as OILC installation representative. 'Anyone being disciplined or pursuing a grievance was entitled to take along a work colleague – and that is what I was. But we were never recognised.' He was working as a helicopter landing officer on Brent Delta. 'We had a good crew of safety reps, a good committee, providing reports and chairing our own safety meetings. Then Ronnie announced he was standing down. I had been a trade union member since I started my time as a plumber, but when I went offshore I was never involved in a union until the disputes in 1989 and 1990, when I was a MSF member. Then I saw the trade unions walk away and I saw the birth of OILC, so I got involved. Then came the opportunity to stay at home. I was a bit hit financially, dropping about eight grand.'

From the operators there was still the traditional resistance to the unions. John Wils was Aberdeen director of UKOOA for seven years from 1996. He has had thirty-six years' experience in the oil and gas industry, first with Shell in the southern gas sector and on the Brent gas systems, then with Hamilton Brothers and latterly with BHP as manager of their Liverpool Bay operation. He came back to Aberdeen in the mid-1990s at a crucial time. The dust had just settled after Brent Spar and public regard for the oil companies was very low, according to a Mori poll. It was an era of change – peak production, smaller discoveries, falling drilling activity and the start of the big mergers. UKOOA were in the process of restructuring and their attitude to the unions had not changed appreciably. 'There was a dialogue, but it was more on generic issues, such as safety and the concerns the unions had. Human relations on platforms were clearly up to individual companies. If people voted for union representation, the union rep had a right to go out to the platform to talk to the workforce. If, as a result, the platform staff took a vote, then the unions could be recognised.

Oilfield labour matters have been the concern of the Offshore Contractors' Association (OCA) since they became a limited company in 1996. There are currently seven full members, with fifty associates. The link between unions and industry management is that they have the powers to negotiate collectively on issues such as rates and conditions. Historically, the conractors came into their own during construction. OCA executive Bill Murray said, 'The biggest change was that as the contractors' capabilities were recognised, the operators were able to allow them further into their business, giving them more responsibility and accountability.'

He explained that they didn't always go along with UKOOA. 'We have a fairly close working relationship, but the two interests are never going to fully coincide. On the commercial side, there is a conflict because obviously the operators want the best price for work and the contractors want the best margins. There are areas where we cooperate – in training and safety, for

Plate 105.
Bill Murray, chief executive of the Offshore Contractors' Association. (OCA)

instance. Similarly, there is a common interest in influencing government on taxation or licensing.' As to who calls the shots offshore, the OCA chief maintained it was much more complex and subtle than simply 'Do what you are told or you don't get the contract. Decisions on contracts are purely commercial. Decisions on operational matters are normally through influence. The client can't dictate. The processes are more open and some of the contractors now are bigger than their clients. Overall, it is a well-regulated but efficiently functioning market place.'

The first suggestion of union recognition was in 1999, when the OCA signed a consultative partnership. The long fight for recognition – but not the union disputes – ended in April 2000, when union rights to the oil and gas workforce were included in the Employment Relations Act 1999. The following month, the two main unions signed an historic industrial relations agreement with the OCA. Amicus have two recognition deals with the drilling contractors' and the floating production operators' associations. Murray said they had 'fought each other to a standstill' during the hook-ups. 'We saw that this was not productive for either party and we've gradually found ways of working together to the benefit not only of the workforce and the companies involved but for the whole industry. We have just set up a lifelong learning partnership on training, with the unions and Robert Gordon University, using social funding never before available. Similarly, on the safety front we hold an annual seminar and we take ideas from that together with the unions through into Step Change where we are both represented.'

Graeme Tran came to Amicus from representing the AEEU at the Faslane nuclear base. 'People believed we had signed sweetheart deals, but the proof is in what you deliver in membership. We are the first-ever union to hold a proper secret postal ballot on a pay offer in the drilling sector where there are six and a half thousand workers. We put out just over 2,000 ballot papers. We also sent out a circular to about 7,000 people offshore. UKOOA do not recognise unions but we liaise with them on the employment practices committee because they want to know the problems. Things have changed in terms of the right to representation for grievances and disciplinary hearings, but there is still reluctance. So what we have to do is to pick off the operators one at a time.'

Steve Harris, communications director, said UKOOA's position continued to be that the majority of the members still did not recognise unions. 'I would guess we do have a relationship with them, in particular with TUC-affiliated unions, and they have more of a voice now. This came, strangely I think, after Brent Spar, when the operators decided talking to the DTI really wasn't enough and they needed to engage with a broader range of stake-

Plate 106.
Graeme Tran, organiser of Amicus. *(Amicus)*

holders including the unions. I suspect also that the change of Government in 1997 had an influence when the unions returned to the table. I also think they were playing a more constructive role in the industry. Now we lobby the Government together around an agreed programme and that is helpful and constructive.'

The exception is the OILC. They have always challenged the exclusivity of the deal in 2000 and were not involved. According to Jake Molloy, the only clause that really matters says that no employer shall pay any less overall value than that set out in this agreement. 'That means they can pay you what they like. These unions were recognised offshore for the purposes of collective bargaining and officials were sent to develop a network of shop stewards, but I think if they are being honest, they failed miserably and they have done more damage than anything else to the ability of unions to establish themselves. They are now seen in the same light as the HSE – in the operators' or the employers' pockets.' In a pamphlet prepared for the OILC, *Union Recognition in Britain's Offshore Oil and Gas Industry* (2003), Professor Charles Woolfson makes the point that, by voluntarily recognising two specific unions only, the OCA agreement virtually negates the provisions of the 1999 Employment Act for any future ballot by the workforce to choose their own unions. So the OILC still operate outside the pale. Their principal tactic remains to challenge individual companies on cases raised by their members, through the courts where necessary.

UKOOA claims that one of the difficulties is what they see as the constant rivalry between Amicus and OILC for membership, with each trying to outdo the other. For the OCA, who don't recognise the OILC, it is all a matter of history. 'When we wanted to improve relationships with the official unions it meant that had be to the exclusion of the OILC. But there were also other unions we de-recognised at the time.' Graeme Tran said it was unfortunate the OILC would not come on board. 'We seem to be fighting against one another and the winners are the operators and the contractors. The OILC call on us to rip up the agreement. That is not going to happen. They have served their purpose to bring people together. I don't see why we should be divided now.'

If it is considered that the unionisation of the workforce is an essential element in the successful provision of health and safety as well as in the maintenance of good industrial relations, then progress towards the necessary sizeable representation offshore is still painfully slow and resisted. Of the 24,000 people employed in the North Sea, a generous estimate would be that there are around two to two and a half thousand acknowledged members in the official and unofficial unions.

Plate 107.
Steve Harris, director of communications, UKOOA. *(UKOOA)*

Women and Home

[12]

'Actually, it surprised me when I got offshore how men would have their slippers in the accommodation. But why should it surprise me, because that was the man's home when he is offshore. I sometimes think that as a visitor I was disturbing something a guy has got used to, a fortnight when he can sort of settle into another home. A bit like a woman in a gentleman's pub, it doesn't quite gel. Sometimes, if there are just one or two women offshore it's disturbing to the guys to have this one oddbod to cope with there.'

The particular 'oddbod' in question is Jane Stirling, who is married with two daughters. She is also the shore-based technology manager for Total E+P, the French oil and gas giant who operated Alwyn, Dunbar and Elgin Franklin fields in the North Sea, and the 86 Fergus Gas terminal onshore. But she began her career offshore as a structural engineer in 1980, on Occidental's Claymore and Piper platforms. She qualifies for the designation of 'a woman in a man's world', one of a few hundred females who chose to work in a 'macho' society – the domain of hard men in hard hats in a hard, physically demanding occupation in a dangerous, unpredictable environment. Or at least that is the conventional view of life offshore.

Women have worked offshore since the early days in the northern North Sea, but not in large numbers. An Inland Revenue survey in 1996 revealed that the number employed was 1.6 per cent – 429 out of a total of 26,850, and that included domestic and medical activities on platforms. Since then there have been no statistics, and in the 2002 Department of Trade and Industry index, the workforce is identified by occupation but not gender. Unlike Norway, where, according to a study by Kathryn Mearns and Rhona Flyn, of the University of Aberdeen,* women now make up 9.5 per cent of the oil industry's official workforce, with the majority working for oil companies – 907 of 5,882 staff – and the rest – 616 out of 1,071 – employed

---

* Mearns, K. and Flin, R., 'Applying science in a man's world: women in science, engineering and technology in the offshore oil industry', in Masson, Mary R. and Simonton, D., eds., *Women and Higher Education, Past, Present and Future* (Aberdeen, 1996).

[223]

THE OILMEN

Plate 108.
Sharon Morris – one of the comparitively small army of women who work offshore in the North Sea. Sharon is operations HSE adviser for Lundin Britain Ltd.

in catering. The researchers calculate that the ratio of women in the UKCS on the 1995 figures of 29,000, ranges from 0.5 per cent (145) to 2.7 per cent (783). Again this includes the more conventional female roles. They do not indicate how many women are working either in science-based occupations such as geology, or the 'heavier trades' such as engineering. Kearns and Flyn say that although both countries have legislation to protect women against job discrimination, Norway 'seems to have actively encouraged' their recruitment 'for both cultural and political reasons'.

Just as elusive as the statistics is any evidence of actual discrimination. Some companies, such as Schlumberger, have a positive policy of recruiting women for these non-traditional roles. Aberdeen-born Euan Baird, former chief executive, once said that they had an opportunity to be 'more aggressive than what we were expecting and what we were achieving on the gender diversity side'. There could be no better example than the current managing director, UK, Ireland and Faroes, for Schlumberger Oilfield Services. Alison Golligher, a graduate in theoretical physics and in petroleum engineering, is the highest-profile woman in the UK oil and gas industry – the only one of her kind. Other major companies, such as Shell and BP, proclaim policies of 'gender diversity' and promote schemes to attract young female science and technology graduates, while similarly encouraging youngsters at school. A very practical reason is the current shortage of skilled people to replace an ageing workforce in an industry not always seen as attractive or environmentally friendly.

Among oilmen, there is a mixed reaction to the presence of women offshore. Former Chevron OIM John Nielsen said he found it 'very interesting' when they first brought women out on to the platforms. 'You might get the odd engineer, but she would be very rare. Women really started in the catering and cleaning side. Over in the Norwegian sector it is really quite different. They do have a community on the rigs and platforms, with more women. In the UK sector you used to have one woman and 250 men. They created a problem initially, but it wasn't their fault, it was the men. We had some good women engineers and a woman drilling rep with Chevron. I really don't mind who does the engineering, but what you had to watch was that some supervisors would spend more time with the women and you don't need that.'

Derek Stewart, a ROV technician on a variety of offshore installations, believes women have gone offshore in a big way. 'There are a lot of women engineers and, of course, your stewardesses. You don't have many British women, and you never get Norwegian women on this side; they are too expensive. What you have are loads of Polish women and Russians. All very hard workers. I think they have civilised life offshore – people tend to shave more. Some guys wouldn't shave for a fortnight.'

Former offshore medic Caroline Russell-Pryce was sometimes very much the lone woman in a man's world at a time when the fact that females worked offshore was relatively unknown. 'I remember answering the phone once. This woman was absolutely horrified because she hadn't realised women worked on the platforms. I think a few men told their wives or partners there were no women.' Caroline, who comes from the west coast, was looking for a challenge as a change from nursing onshore. 'My father travelled extensively when he was young and I think part of that was in me; I needed to do something different. It wouldn't have suited everybody.' She did a week's trial on the rig *Safe Reporter*, on a hook-up on an Amoco platform. There were 700 men on board and she was one of two medics. 'I enjoyed it. With that amount of men it was very busy, and being a construction job there were a lot of cases of debris in the eyes and so on.' Caroline had paid for survival training herself; for the medics' course she was helped by Moray Enterprise. She was thirty-six and fit and she enjoyed the three-day RGIT course. On her first trip for Offshore Medical Support, she was the only woman on the helicopter. 'The worst thing was that everybody else seemed to know what they were doing. I only knew where I was landing because I could see the name on the heli-deck.' Some trips were eventful. 'One time helicopter fuel started pouring out of the ceiling. It was a bit alarming. Another time there was poor visibility and we could only see the platform on the radar. Suddenly it loomed out right ahead of us. He had to go almost straight up beside the platform to avoid it.'

THE OILMEN

Plate 109.
Offshore husband and wife team – former medic Caroline and Steve Russell-Pryce who met up on a drilling rig.

One big change was in her pay packet. 'When I came out of nursing, I was on about £15,000. As a medic I started on £24,000. Latterly I was on £32,000 when I finished on the Dolphin rig. I'm back now on just under £19,000 as a senior staff nurse in the emergency theatre. When I started there were a few stewardesses, but I came across one female radio operator and one female engineer. There were other medics, but not as many as males.' Her first installation was a floating storage unit between the Fulmar and Auk fields, but she moved around. 'You were always the only medic unless there were more than 500 on board. You trained the first-aiders as helpers if something went wrong. One day we had a death – a guy had collapsed in a phone booth. We started resuscitation and then I got a computer printout of the phone calls and established when the phone had been disconnected. It had been too long and he wasn't viable, so I discontinued resuscitation. The procurator fiscal supported that. Otherwise there were always crush accidents, particularly on the drilling rigs, hands and toes. One guy fell and fractured his spine. I packed him full of painkillers and sent him onshore. But you never heard what happened to them afterwards, so it was pretty much emergency medicine. Or they would have been somewhere they shouldn't and picked up something they couldn't take home to their wives and I had to dispense antibiotics. Medics also had to do the catering inspections and look after the total health on the installation. Living on a platform for two weeks, you don't want somebody getting food poisoning.'

The practical no-nonsense nurse said the men weren't as macho as they liked to make out. 'They treated me reasonably well. Some women made the mistake of trying to be one of the boys, swearing and behaving like them. I found you earned much more respect keeping yourself slightly separate. I didn't mix with them in the evenings, and onshore I kept my social life completely separate.' Caroline didn't maintain her professional reserve all of the time. She met her husband Steve on her last rig, the *Dolphin*, where he was radio operator. She was eventually made redundant and went back to hospital nursing, while Steve retired. They now live in Inverness. 'I'm glad I did it – the offshore work. It was an excellent experience which gave me great confidence to deal with anything.'

When young Sarah Wingrove went on a drilling rig as an assistant to the derrickman, he looked at her. 'I have three daughters, a female cat and a wife – and now when I come to work I have got you!' That was the kind of reaction the female mechanical engineer, one of the recent new wave of young, educated women to join the oil and gas industry, had to face. But engineering was in her blood. She had emulated her father by graduating from university with an engineering degree in 1996. 'There were about 125 in my third year and only a handful – six or seven – were women. I guess engineering is not the kind of work women would traditionally go into.'

Plate 110.
In full working and safety gear, engineer Sarah Wingrove on helideck duty on Shell's Rig 135, in the winter of 1997.

THE OILMEN

Sarah had spent a month with Schlumberger on work experience and she had also attended a lecture about the industry. 'So I knew a little bit, but not a lot. It was mostly oil companies I applied to. They certainly paid well; you could travel with your job and it was quite a varied career with lots of different avenues. I had quite a few interviews, but in the end Santa Fe hired me.'

Sarah's first posting was to a semi-submersible on Forties. 'The reason the first trip sticks in my mind was because I was choppered off early. I had hurt my finger twelve days into the first hitch.' Santa Fe did have other women on their books, but the 22-year-old trainee was the only one on that rig at that time in that capacity. 'I suppose it was quite exciting, the first few trips, until it became mundane. I thought, "Do people really do this – spend all morning scrubbing a deck which is going to get dirty in six hours' time?" It was an eye-opener. We did a lot of scrubbing, a lot of painting, the preparation of bulkheads, and I worked the cranes as well. And, yes, I was treated differently, being female and being quite a bit younger than a lot of my colleagues, especially the roustabouts, who had had other careers before, so they were probably late twenties or older. It definitely is a macho environment, but it depended on the guys; some would help you out, some too much; some would have respect for you and want you to get on and others really didn't want you to be there. The accommodation was in two-man cabins and I shared with my opposite number. That was fine; no problems there.'

Sarah eventually did ten trips roustabouting and then worked on an upgrade project on the Santa Fe 140 rig in a shipyard, driving a forklift truck. 'I also worked on the blowout preventer, which was quite interesting. Then I went out as a roughneck for maybe two years.' The young woman is slim and hardly the build of a roughneck. 'I am fit but not necessarily strong. I don't have the weight a lot of the guys have, but 95 per cent of the job I could do as well as anyone, although I had problems throwing tongs – these days they use "iron roughnecks". Rig 140 had a cab up at derrick level with an arm you had to get out on, but I had inertia reels and harnesses, so I was reasonably safe. The equipment was there but you had to insist people used it. We worked right through the weather and there were appalling spells when we did nothing but scrub and paint because the rig couldn't operate. Every time I went back to the beach it was, "When am I going to get out, just another trip and just another trip?" You can only put up with that for so long.' At that time, 1999, the oil industry had slipped into a lengthy recession and derrickmen had to work as roughnecks and roustabouts. 'There was talk about my being fast-tracked to driller. But it wasn't going to happen in that environment.' So she applied to Shell but was unable to get a job and went to work for six months as a drilling engineer with BP. 'They were excellent for training. Working on the South Everest field, in the central North Sea, there were plenty people to explain things and point you in the right direction. I was quite happy there.' Then Shell offered her a job and she has since been working on wells for the company's massive £350-million development of the Penguin clusters, near the Brent field, involving subsea completions and tie-backs from five fields.

Sarah started on about £15,000 a year, but it increased rapidly. 'It is still a well-paid industry as far as engineers are concerned. There aren't many jobs you get more than £40,000 without ten years' experience.' Initially, her parents were worried about her, but she said they had come to terms with it now. 'I think offshore you have to be able to give your worth. I don't think allowances should be made for you because you are female. I don't feel guilty; it shouldn't have been male dominated in the first place. I am just helping to redress the balance.'

When Jane Stirling joined the construction industry in 1976, she wanted to pursue an interest in structural engineering. She had been working for Wimpey Construction on clamping technology in the North Sea, where many installations had been built to withstand the wave loading more typical of the Gulf of Mexico rather than that of the North Sea. The result often became evident in the splash zone where fatigue cracking can occur. Jane came north in 1982 with Gordon, her husband, also an engineer, when Wimpey set up as an independent offshore firm. Gordon was already working for Omisco partly owned by Wimpey. Jane later moved in 1984 to Occidental, whose platforms had begun to develop fatigue problems. Her knowledge on clamping won her a job as a junior engineer. The platforms were Piper and Claymore.

As a woman engineer she was something of a rarity in those days. 'I think it was a novelty for them and a novelty for me as well.'

She remembers her first trip offshore. 'I was sitting at the heliport and the safety induction film came on. I was horrified because it indicated that if you needed to relieve yourself on the helicopter there was a bottle provided. I could imagine all these men passing round the bottle. I announced that if anyone wanted to go, could they please go now.' She spent more time on Piper than Claymore. 'It was a friendlier platform for a woman, because it had one- or two-man cabins as opposed to more typical four-man cabins. On Claymore you had to walk through urinals to get to the cubicles.' Jane, who now works for Total, is surprised that there still aren't many women engineers offshore. 'Women haven't chosen to go into the engineering professional fields in a lot of cases. Those who have gone into engineering don't sustain the course. We have a few, but you would think it would be like accountancy; twenty years ago it was male only and now it's about fifty-fifty. We still haven't got rid of the stigma that for engineering you must be good at maths, and many girls don't enjoy maths.' Jane acknowledges that for some women, engineering can be a difficult choice. 'Unless you actually want to do it, you are not going to get through. There have been countless times when I was working on construction sites when I've thought, "What am I doing here?" The physical side has never worried me. Fortunately, I married an engineer, and I understand his pressures and he understands mine. Neither of our daughters wants to be an engineer, which is not a bit surprising.'

Schlumberger regional managing director Alison Golligher only arrived in the North Sea three years ago, but she has already made a considerable impact. She is quite adamant that it isn't purely a man's world. 'I prefer to think that it is anybody's world – an adventurer's world. People join the industry because it is different. It encompasses traditional skills and things at the edge of a lot of people's comfort zones.' She had been keen to work for BP 'on the travel side, which was the operations engineering side'. But she was told she didn't have the necessary engineering qualifications and her science degree from Edinburgh was not suitable. Alison was then sponsored by Shell for a engineering masters degree at Heriot Watt. She joined Schlumberger in 1988 and went to Brunei as a field engineer. She was the only woman candidate interviewed and the only one on the training course in the Far East. 'Obviously, nobody goes to university for a wire logging degree, so all our people are trained from scratch. In those days it was a thirteen-week, 24-hour intensive training schedule – we used to call it "Boot Camp". I was always treated fairly – spoiled rotten in fact. I was lucky; in Brunei, people are very kind by nature and they were curious and respectful. There was no "Prove it." It was, "This is what you want to do and we will work as hard for you as for the next guy."'

Apart from a spell in the Norwegian sector, Alison had never worked in the North Sea, where she thinks it would have been more difficult to learn her business. 'I would have worked alongside engineers who built their own cars, whereas I didn't know how to screw things together. Although I had all the theory and the technical side, I learned the practical side in a relatively peaceful environment where people were more forgiving. I would have had a lot harder ribbing here. It depends on your personality as to whether you can cope or not. I like to think I would have done.' In the jungles of Indonesia, working in remote locations for long periods, conditions were totally different from offshore. 'When you have been up three days, absolutely filthy, with the same clothes on, driving home and the truck falls into the ditch,

Plate 111.
Schlumberger's regional managing director Alison Golligher – highest-ranked woman in the North Sea oil and gas industry.
(Kate Sutherland)

there is no recovery vehicle, so you get your spade and start digging. Then you think, "I went to university for five years to do this?" But it was a lot of fun as well.' On one jungle journey, she and three colleagues saw what they thought was a bear sitting in the middle of the road. Instead, it turned out to be a fully grown female orang-utan. 'We stayed absolutely still. The animal stared back, then shrugged and ambled back into the jungle.'

She believes that women take a different perspective from men. 'Women tend to be – as my team tell me – very good at doing lots of things at once. So we are good at multi-tasking and good at selecting what to do, because we are expected to juggle lots of things, homes, families, careers, and it seems to come more naturally. Some women also have a more relationship-orientated look at things, or at least that's what all the books tell you. I am often told I tend to look at things by seeing things from the other person's point of view.' Alison worked in America, Norway and France in a variety of posts, and before coming to Aberdeen she was a vice-president in Paris overseeing new technology. At her level, there are twenty-seven regional managing directors, and she is one of three or four women. In the UK all her peers are male. 'I am unique in this part of the world, but we have seen a big effort in the industry and I have met some extremely capable ladies in my short time here. Shell and BP have some very senior ladies, and a lot of companies with enough critical mass and size are really working hard towards that. You don't see very many female drillers, but these things may take one or two generations, unless you recruit 50 per cent women on your intake. In fact, we are at 25 per cent in our uptake for engineers, while the company ratio is probably 12 or 15 per cent, and I would guesstimate perhaps 8 per cent across the industry.' Schlumberger has had an equal opportunity policy for the past thirty or forty years, and that is in national as well as gender diversity. 'There is a very close correlation. I think that having a broader view is useful; lots of people have slightly different perspectives.' Her company, like BP and Shell, works to encourage youngsters, sponsoring the annual Aberdeen Techfest. Alison was also behind a programme at Heriot Watt University to train two women engineers annually. 'I am aware that some folks in my own company are proud of what I do, which is obviously flattering and wonderful. But what is very important is that people respect you for what you do, not for who you are or where you come from, or what gender or race you are.'

In the industry the other important part of the male–female equation is the pressures of the enforced work division on family relations – each different world with its own particular pressures on the men and their wives and partners. As a medic, Caroline Russell-Pryce often had to play surrogate 'agony aunt' for men trying to come to terms with their isolation offshore. 'Some wanted to talk about their home lives, wives or partners, and because I kept myself separate, they seemed to assume I wouldn't talk about these things to anybody else. I think that was quite important out there. The life was quite difficult for some of them. I don't mean the ones who went wild, but just generally because of the work pattern.'

Such a fractured lifestyle is nothing new for the local north-east communities. In Aberdeen, which used to be the UK's premier white fish port, the pattern of trawl and seine-net fishing meant that men worked away from home, spending only short periods with their families, while responsibility for home and children bore heavily on the womenfolk. Both industries – oil and fishing – created similar domestic problems. Archie Robb, an Aberdeen social worker in the 1970s, said that they were also very different cultures, 'With the fisher folk it was a traditional way of life. The women knew their men would be away a lot and it was pretty much a matriarchal society, very close-knit. The oil wives were more isolated, many unused to a totally new lifestyle. Suddenly, the husband was offshore and the wife had to adjust; most did, but some didn't. Being on their own a lot meant there were temptations onshore, which caused a lot of difficulties. Alcohol also posed problems for the men coming ashore.'

The bulk of the new labour force was recruited from central Scotland, the north and Midlands of England. For these shipyard workers and factory hands from close communities, their new lifestyle was alien and disturbing, as Ronnie McDonald saw. 'If the local guys were generally settled, with wife and kids, they got used to the routine. They were content that things were being looked after at home. There was a regular wage and some of them made positive use of their time off. But a fair proportion couldn't handle the time off. The bevy, for example, leaving the single guys out of it – a lot of marriages couldn't take it. It was the fortnight off, disposable income, time on their hands, the pub. The vast majority got a lot of benefit out of offshore work; they saw more of their mates than of their wives and the whole problem maybe came into sharper relief.' Ronnie added that child maintenance was always a difficult issue among the offshore men. As an oil constituency MP, Malcolm Bruce of Gordon deals with oil families. 'To be fair, they don't come with problems that are different from any other constituents, but there may be a higher divorce and separation rate because of the life they live. They probably seek more help with Child Support Agency problems and there are obviously tax issues.' At one point, Aberdeen had the highest divorce rate in the UK – 7.5 per cent compared to an average of 6.4 per cent – but welfare workers are reluctant to blame it on the oil industry. One oilman claimed his marriage had lasted thirty years 'only because he had accepted the house was his wife's domain'.

Graeme Paterson, a veteran of the industry, now feels it is getting progressively harder to go offshore. 'The only thing that's keeping me offshore now is the pay I'm making. It is a wrench every fortnight. You think, "Oh, here we go again. I've got to go away." You miss a lot. That last trip, I wasn't there for my girls' eighteenth birthday. Nowadays I am getting too old – I am in my fifties. I sit waiting for a taxi and you sometimes think about cancelling it. I'm not going. Then you think, "What else am I going to do, what am I going to live on?" It's been an experience and it's been good to me. It's given me a lot of things I would never have had, a nice big house. But it is getting harder.'

Brian Porter worked away from home for thirty years on contracts in the pipe-laying and coating sector. He admitted his family life had suffered. 'The wife just about brought the kids up on her own. A lot of marriages didn't survive because of it, a helluva lot. I know blokes who have gone home, the wife has cleaned out their bank account and had a bloke living with her. It devastated a lot of them, although to be honest some deserved it. They never sent money home and they were always away, just for the freedom. They were always drunk and spent all their money. But the genuine ones were doing it for their families. It was a shame for them.' The wilder days of the oilmen have calmed down now, according to former Chevron OIM Alex Riddell. 'They are more mature and responsible these days. The construction guys used to be the funniest. You got phone calls from their so-called wives trying to find out where they were. So you had to play it cool. A lot of these guys were leading two lives, saying they were out there for a month when it was only two weeks and they were disappearing elsewhere.' Sometimes McDermott company rep Ken Macdonald had to assess the veracity of some of the hard-luck tales, many of which contained the most amazing coincidences. 'Every time there was a Rangers–Celtic game somebody wanted ashore. The guy's granny had died – again. It used to cost a lot of money. They used to call it "third weekenditis". Once, this guy, a great big rigger, told me, "The wife's cat has been run over." I said, "You want me to mobilise the chopper" – this was the 1970s and it cost £400 – "plus the fixed-wing, for a cat?" I won't tell you what else I said to him!'

But there were of course genuine cases. OIM Andy Lawrie was on the Bruce platform during construction. 'The son of one of the men had been in a car accident in Glasgow. This was midnight and we had no helicopters. I phoned the hospital and was told the son wouldn't survive the night. Fortunately, there was an infield helicopter sitting in the Forties field and I asked for it to be sent. Then I asked Dyce for a helicopter to take the man directly to Glasgow. His son actually lasted a week before he died. Somebody reported to management with half a story and they asked why I had authorised the helicopters when the lad was already dead. But I was able to tell them he wasn't dead at that point. Not another word was said. In the eyes of the men, BP were the best company in the North Sea. What did it cost? That man needed to be there.'

As Ronnie McDonald said, marital casualties were high. Jake Molloy had married in 1980, the year he first went offshore. 'Being away from home to work wasn't new, because I had worked all over Scotland during my apprenticeship. I wasn't married then and didn't have any kids. In 1981, my wife had our son and things got a wee bit harder then. My wife was a medical secretary, but she wanted to stay at home. I was earning reasonable money, and that made things a bit easier. Then she wanted to go back to work and when that happened she wanted me at home to share the load. By that time I had become indoctrinated into the offshore game and I felt to come back onshore at that time probably wasn't the best thing to do.'

Shell driller Martin Reekie's marriage was another casualty. 'Like most folk in the industry, I managed to get married and divorced fairly quickly. There aren't many I have met who haven't been in that situation. Whether it's because of the job, the money or anything else, I have no idea. It is quite scary, the number of people I know who did get married and divorced rapidly.' The lifestyle of a crewman on a supply boat didn't suit Ian Sutherland's wife. When they married, she had decided he was away from home too long. 'I heard of lads who would get into a card game before they were due home and blow all their money. With others it would be, "Do you fancy a quick pint?" Ten days later they would turn up at their doors.' One of the key questions, Jim Souter, another former BP OIM always put to new recruits was, 'How is your wife going to take to this?' 'If I got the answer, "It's nothing to do with her," we would find somebody else, because you knew there would be a problem. As one of the guys said down in Grangemouth, "Every two weeks you dry out and every two weeks there's another honeymoon."'

Compared with working in the oil construction yards, the lifestyle away from home suited Australian Ralph Stokes, who worked offshore as a welder. 'But for others it had its problems. When they got off, all their mates were working. So there you were, sitting in the house and unless you had a hobby you would end up going down the pub through the day; unfortunately that is what happens with a lot of oil people. Luckily, I had a croft to keep me busy and my wife didn't work through the day. So that was okay.' When Dunecht joiner John Selbie first went on to *Staflo* to roustabout, he injured his leg and had to be evacuated. 'My wife said, "You shouldn't be out there, min, you are a jiner, you're nae an ile worker." I went back three weeks later and the boss said, "I didn't think I would see you again." It was a struggle for the first three months. I hated it. I used to lie in my bunk thinking about the kids; but you can't exist out there like that. Then I learned. I must have got to know the job better and time just flew past. You would switch back when you were ready to come home and it was a case of "Now, what is happening at home?" My wife wasn't too happy at first and she didn't keep well. She told the doctor she was a bit stressed when I was offshore. The doctor said, "That could be causing it." Then she passed her driving test, which gave her mobility, and she was fine.'

One of the North Sea pioneers who came north on *Sea Quest*, Joe Dobbs, is convinced that working in the industry made his marriage. 'People who don't go away don't get that great feeling of going home. We used to come back the quickest way we could; we hired cars, we flew, we caught sleepers.' His great mate, Swede Lingard, was of a similar view. 'I met my wife two weeks before I went to work on the rigs, anyhow. I always used to come back to Lincolnshire and we got married in 1970. I have always come straight home even when I worked all over the world. Before I retired, I was in Colombia in South America, three weeks on and three off, and I always got on the plane a bit quick. It helps if your wife doesn't know anything else. Anyone from an ordinary job – a bus driver or a mechanic – used to have a lot of problems.' The Russell-Pryce husband and wife team were a novelty in the oilfields. Caroline's husband, Steve, explained, 'She actually took my job

THE OILMEN

on the rig when I became the radio operator and after that we always worked together offshore. We were friends for about two years before we got married. There was nothing untoward, although she was a bit of all right, but the boys on the rig never believed us.'

Shell's Mike Waller worked on some of the high-pressure, high-temperature well programmes offshore and was on constant call. 'What comes out of all this is the credit that's due to our wives. Certainly it is the case for me, and I know for other guys as well. The women behind the men in the oil industry are sometimes not appreciated.' A number of companies, though, realised the importance of women in the special circumstances of the offshore pattern and flew parties of wives out to the platforms to see for themselves their menfolks' uncommon workplaces. Another support for onshore wives is the Petroleum Women's Club of Scotland, based in Aberdeen. Founded in 1972 by a group of five oil workers' wives, the club, which carries out work for charity, now has 300 members of forty different nationalities.

One oilman's wife who decided to act independently to deal with the stresses and strains on 'offshore' families was Gina Sims. Following the Chinook helicopter crash in 1986, she started a wives' support group in at her home in Polmont, West Lothian, and raised money for the disaster appeal. When a number of men from her area died in the Piper Alpha tragedy, they started a formal local wives' group. In 1992 the Offshore Wives Link Support (OWLS) was officially launched, with Lady Cullen, wife of the inquiry judge, as patron and with support from a number of oil companies. The OWLS network's main activity was manning a 24-hour helpline for wives. Sadly, it closed down through lack of funds. Nothing has taken its place.

Camaraderie and the Crack [13]

Any veteran oilman who has spent time in Shell's Brent field in the East of Shetland Basin within sight of the Norwegian sector is liable to recount the story of the mythical 'Statfiord disco'. There are several versions of this famous 'leg-pull', and Bristow Helicopter operations manager Richard Enoch describes the version he himself played when he used to fly out of the Northern Isles.

'It was called the Stafiord disco because the story went that helicopters were regularly available to take the guys to the Saturday disco on the Norwegian field. And, of course, everyone knew there were women on the Norwegian rigs but not in the UK sector. This guy asked me, "How do you get a chance to go?" I said, "There's a raffle." On the so-called day of the raffle the man was onshore. When he came back, they told him, "What a shame. You'll never guess! Your name came out of the hat and you missed it. So they had to go without you." He was furious.' Rig painter Graeme Paterson verified the story. 'Brent used to be notorious for the number of times guys stood on the heli-deck on a Saturday night, all dressed up, waiting

Plate 112. Stafiord across the meridian in the Norwegian sector – gig for the legendary disco and the fabled women oil workers. *(Øyrind Hagen, Statoil)*

for the chopper to take them to the Statfiord disco.'

Such classic leg-pulls, along with the wind-ups, the parade of crazy characters, the rough, iconoclastic humour and, above all, the 'crack', represent the totally unexpected, largely unpublicised side of life offshore in the North Sea, at odds with the popular image of a tough, highly pressurised and hazardous existence. 'It's because of that environment there's so much humour and leg-pulling,' said Graeme. 'There's a lot of camaraderie and it is a relief for the guys. If you didn't have humour you would just get depressed and want to go home.' Joe Dobbs has retained the friendships he struck on *Sea Quest* nearly forty years ago. 'It was just comradeship. Obviously, you can't get on with everybody. But I never saw any trouble out there. And we used to get a couple of cans of beer – the only rig that allowed that – because it belonged to BP Tankers, but nobody abused it.' Shell toolpusher Mike Waller said that once they were more expert they built relationships and became part of a team. 'There was a great camaraderie, to such an extent that if you were moved on to another crew, you didn't like it.' Former Chevron OIM John Nielsen thought the early days were a lot more easy-going. 'Then it all got a bit uptight, a bit more intense, with these deadly serious young engineers. We had quite a lot of light-hearted fooling around and you got a lot of wind-ups. It never did any harm.'

The Classic Wind Ups

Some of the wind-ups are classics. When Energy Minister George Younger was on an official visit to Chevron's Ninian Central, the crew were ordered to spruce up the installation – and themselves. Several of the men were given new overalls, hard hats and boots. The entire walkway and the heli-deck were painted yellow. The contractor workforce pitched in to help. During his tour the Minister entered a huge module with banks of electrical panels and noticed that one of the crew had a little windmill sticking out of the top of his hard hat. 'That's very innovative,' he said to the man, 'does it work?' The man, who came from Elgin, replied, 'Oh, yes – watch.' And he ran from one end of the module to the other with his wee windmill spinning. 'That was the last time we saw that guy, needless to say,' laughed Jake Molloy, who was on the platform that day. 'But it was that kind of thing that made life bearable.'

Jake maintained that that Moray man made it his business to make the American management's lives hell and he thinks he also performed another mad stunt, although it was never proven that he was in fact the culprit. 'New rubberised flooring laid in the workshop had been allowed twelve hours to set. Overnight, someone got into this huge area and left a bare footprint right in the middle. Now, it was unreachable other than by the overhead crane. So this person had obviously hooked himself on, taken control, stuck his foot down and come back up. The guy who laid the floor said that whoever had done this would have burned his foot, because the flooring was corrosive. Funnily enough this Elgin guy did walk with a limp for several days after.'

The humour could also be cruel. Like the joker from Elgin, the diving crew on a gas-boosting concrete storage platform midway between Norway and the Buchan coast during the 1970s seized every opportunity to annoy their American colleagues. Among them was Frenchman Michel Euillet. 'We had a rather strange American diving rep from the Oxy school of diving – Occidental already had an extremely bad name. He was immediately recognisable; first he was American; secondly, he was a "coonass", and thirdly, he had a gold Rolex with diamonds which he always complained cost him a fortune in shirt cuffs. 'One day one of the divers had speared a monkfish which had a knife inside it. The American took the creature up and said he used to spear fish like it. The fish's mouth was wide open and, I don't know why, but he put his hand inside. The monkfish went snap and locked on his wrist. We said, "Well, you can always use the knife inside to cut off your hand." And we left him. Eventually we went back and freed him. We never saw him again.'

The offshore crews used to go to fantastic lengths to trick their victims. A new young Lloyds representative on Ninian North, who was quite naïve, had a portakabin office on the lowest deck of the platform. John Nielsen said, 'We were discussing the legendary hundred-year wave and how the platform had been built so the lowest deck would be above the level of this wave. The Lloyds representative wondered if the north deck was high enough. We said that it could well be below. Then he asked when the wave would come, and the engineer said, "You get them round the clock, then you don't get one for a long time." One absolutely beautiful day – flat calm sea – he went to his office. They had it all set up, scaffold poles under his portakabin, and hoses rigged. Then somebody ran past banging on the windows, shouting, "The hundred-year wave is coming, the hundred-year wave is coming!" Then they started rocking the portakabin and spraying it with the hoses. And this bugger – scared to death, he was – came hammering out of his office, straight out into this beautiful sunny day.'

In the early days on the Brent, Peter Carson said they used to play all sorts of stunts on the new boys. 'You would get a young engineer, straight out of Cambridge or Oxford. And he would be told that his billet was on 34C, third floor of the Brent, room 4, bunk C. And he would be warned,

THE OILMEN

"Don't put the lights on, there are people sleeping." He walks in and sees somebody is in his bed. Back he comes up the stair and says, "There is somebody in my bed." One of the guys says, "Hey, we can't have that," and he takes a baseball bat, throws on the lights and then cracks the bat on the head of the guy in the bed, shouting, "Let that be a lesson to you. Never sleep in the wrong bunk." Of course, it was one of those blow-up dummies.'

Jake Molloy learned early on about those rubber replacements. 'When I first went on board Ninian Central, there was all the usual leg pulling. On that first day, I was instructed to see the camp boss. I was supposed to ask him about rubber dolls to be repaired. I said, "What's this rubber dolls business?" "Oh," said the camp boss, "the guys can hire them at night. Then afterwards they lie back and have a cigarette and the dolls get burned. So you just get a puncture outfit and put a patch on them." Aye, right. After that you began to realise the amount of leg pulling – a bucket full of boltholes, go for a long stand – crazy, crazy old days.'

Two crewmen who worked for Ocean Inchcape were also at the centre of a wind-up involving a dummy. They were called Willy and George, and according to Ian Sutherland, who was working on the supply boats at the time, their many mad exploits were legendary. 'One time they spent the entire six hours' sailing time out to a rig making a dummy out of an old boiler suit and an old hard hat with a plastic bag filled with tomato sauce for a head. They tied the dummy to a winch, then they got out on to the deck and signalled to the dummy, which, of course, was not going to do anything. The crane driver was watching all this. Suddenly one of the boys threw down his hard hat and began jumping up and down. The next thing is he grabs the fire axe, runs up to the winch and decapitates the dummy. Red liquid everywhere. Then the two of them picked up the remains of the dummy and threw it into the skip. Apparently the crane driver had to take a break for 15 to 20 minutes. Then he called up the captain and said, "I think one of your guys has been murdered."'

When Roy Wilson, a young 'green' graduate engineer with Arco, was on his first well he had to undergo a favourite initiation ceremony instigated by the macho American drillers. 'This guy said that as I was the young engineer, the company man, I should light the well. "We'll give you a gun that fires roman candles and you've got to climb to the top of the derrick and fire the candles into the gas." Of course, the noise was terrible and for a young guy not used to climbing the derrick, going up there with this roaring gas, it was frightening. But I was man enough for it, if that was what they wanted me to do. So I got in the harness and I was starting up and they said, "No, no." They were only joking. But that is how they flared the gas. When it caught, it went bang and the heat was phenomenal and the noise – you couldn't hear yourself. Of course, they just killed themselves laughing at me.'

On Chevron's Ninian North, John Nielsen and his back-to-back used to instigate a Sunday morning clean-up. 'Everybody was involved, including the OIMs. They were all given a black plastic bag and had to go round picking stuff up in the areas they were allotted. There was this scaffolding crew out – four of them. One was a bit naïve. They were told to clean up round this corner. They came back five minutes later, staggering underneath these bags and took them to the skip. The naïve one said, "Give us another bag." He did this three times, until the old foreman said, "That's strange. I never thought there was that amount of rubbish." So he looked round the corner and there was the guy blowing up the black bags.'

The fun and games didn't stop once the crews went onshore. Former BP OIM Jim Souter said the drilling crews had real team spirit – even on the beach. 'Some of the things they did when they hit the bars. One night they were all heading into the Criterion Bar, acting as Snow White and the Seven Dwarfs. The lead guy went in on his knees – "Hi ho, Hi Ho" – without realising that the rest hadn't followed. Another time they went into one of the nightclubs. Jeans and denims weren't allowed. So the one who was wearing trousers was told to go in to the toilet and pass them out the window and somebody else would put them on, and so on. When he passed them out the window, his mates had all disappeared.'

The Characters

Among the thousands of men who worked in the industry there have been some memorable characters. Swede Lingard noticed on *Sea Quest* that the toolpusher had two sets of handcuffs locked in the safe along with the codebook. 'I found out what for later. We had knocked off for breakfast. Now the cook was an ex-boxer, but he could make bloody good cake. Anyway, the toolpusher came in and complained the grapefruit was cut down instead of across. That must have tipped the chef over. He threw a big can of beans at him and knocked him clean out. Over the counter he came with a bloody great meat cleaver and soon cleared the mess hall. Then four of us jumped him and handcuffed him. We should have really twigged, because he went for that same toolpusher some weeks back.'

Shell's Mike Marray remembers another larger-than-life character. 'In the toolpusher's office on this rig there was a large table with two big radios on it – old-style Marconis, with crystals and two channels. Up in the radio room

you had sets with about eighty channels and there was also a satellite link. Anyway these big radios were always on together because they were so unreliable, but they had both broken down. The operator said, "The radio is buggered again." I said, "How buggered is it?" And he picks the thing off the table, chucks it on the floor and it crashes. "Completely buggered," he said.'

On occasions wind-ups had absolutely no effect on the offshore 'Desperate Dans'. Sandy Clow tells of a big ex-farmer from the north. 'He could eat and it was nothing for him to ask for three steaks. He'd eat them and have another one. One time, he went to get another steak and they covered it with Tabasco – the hot sauce. He just ate it, he never batted an eyelid. We were waiting for him to splutter and cough. But he ate the bloody lot – amazing.'

Alan Higgins has a fund of stories from the oilfields. 'The Norwegian skipper on the semi-submersible off Ninian rang to say he had a slight problem. Here's a guy walking up and down outside the helicopter office, waiting to go ashore. And he was walking his dog. It was actually a cornflakes box on a rope, but he was walking up and down, talking to it. He had obviously just had it. So we got him on the helicopter with the medic and he was taken away. Then there was the man who walked into my office in the very early days and asked for a job. I said, "Wait a minute. Where have you come from?" He said, "Aberdeen." In the chaos of the hook-up, this guy had evaded everything; he had no job, but somehow he got out to the platform and arrived in my office. I just had to take him on for his initiative.

'Another day, a colleague and I were walking round the platform and we met this guy in a survival suit wandering about. We said, "What are you doing?" He said, "I am looking for the camp boss. I need some sugar." I said, "Why are you in a survival suit and where have you come from?" He said, "I am off the standby boat. We have run out of sugar and I have just come up the leg." He went back the same way, but we decided, that's it – you can't have people climbing up the legs. Another time a steward was acting strangely. Again it was a medic job. We got him calmed down and put him on the chopper. Two weeks later we got a postcard from Sweden. All it said was, "Who's mad now?"'

Roger Ramshaw of Conoco also has a fund of tales about a Texan toolpusher called Bob Bodie. 'Basically, he managed to rub everybody up the wrong way and he had an absolutely foul mouth. I remember once on the Dundee Kingsnorth rig, he had criticised the eggs at breakfast and then thrown them back. Then he went to write the morning report. I used to pick up the report and when I got to his office I heard a commotion and there was the chef laying into Bodie, who was on the floor, kicking the shit out of

him. This guy Bodie had chosen to bad-mouth, a Glaswegian, had boxed for Scotland and he had been up for assault. Bob was incredibly embarrassed and they took him off the rig that day.'

Alan Higgins recalls a morning in 1980. 'Two guys were sitting on a platform down near the water and they were fishing. All of a sudden a man in a survival suit flashed past into the sea. Big splash. He swam towards them, climbed out and said, "I was just testing my survival suit." He climbed back up the leg and disappeared. A hundred or so feet. I don't know how he didn't kill himself.'

The man who set up the RGIT survival training course, Joe Cross, tells this story. 'A guy who had just been through the initial theoretical training came up to me and said, "I don't really feel I can do the wet life raft." I said, "What's wrong, can you not swim?" He said, "Oh, I can swim all right. I'm a good swimmer, but I have only one leg." I looked at him and said, "But what about the medical examination?" He said, "The doctor just patted me on my thighs and counted one–two." And this guy was working offshore. I said, "I can't in all honesty sign a certificate. You come back at lunchtime and I will take a full drill – but come with your overalls on." It wasn't up to me to make sure he was medically fit.'

Mike Marray of Shell was involved in Joe's offshore survival training at Aberdeen harbour. 'We were all in survival suits and you would go up and down in the davits and lower boats and sail around. Then the instructor would put you outside the harbour, where it was choppier. So he says, "You have to assume this is the real thing. You have just come off a rig, so if you feel sick," he says, "undo your survival suit and you make sure you are sick inside it. Because two days later you might be hungry."'

One of the OIMs was walking the upper deck on Ninian North on a brilliantly sunny day when suddenly a tinkle of water rained on his hard hat. Looking up he saw the monkey board man having a pee halfway up the derrick. Needless to say he was not best pleased and had words with the rig superintendent. John Nielsen said that by the end of the day the platform's cartoonist had pinned a cartoon on the notice board showing the rig super jumping up and down in front of the drilling derrick shouting, "Who pissed on my Derrick?" The OIM's name was Derek. But the good thing was that a 'pig's ear' and pipe was fitted to all the monkey boards to save the man coming down for a pee. When this story was relayed round the platforms, one of the guys said the OIM's first mistake on a fine day was to look up. His second and biggest mistake was to lick his lips!'

There was always more than oil on the seabed of the North Sea, as inquisitive divers used to find. Dick Winchester piloted a manned

submersible. 'Another chap training pilots found the wreck of a freighter sitting on the bottom, in 400 or 500 feet of water, fairly intact. We came up with this wonderful plan to blow the propellers off. Then a Norwegian fishing boat came by and the skipper said the ship was a German freighter from World War Two. And it was full of explosives. So we decided to give up our plan to blast off the props.'

The Rogues

Over-indulgence in alcohol was a contributory factor in some dramatic incidents in the flights Schlumberger's Neil Ferguson made while working on SEDCO 700 off the Irish coast. 'The contract was run out of Aberdeen and there was a chartered flight, an Air Ecosse Bandaraike twin prop to Shannon. They flew out the drill crew, the regular services hands, the cementers, the mud engineers – we would all travel down together. Because it was so far away, they needed a trainee cementer on board all the time. You flew into Shannon, but they wouldn't file a flight plan to Aberdeen until the helicopter had landed. So you ended up spending a couple of hours waiting for a slot and everybody bought booze from the duty free. And then into the bar – fourteen guys all trying to buy a drink. By the time you got on the plane, everybody would be half-cut. The plane has a very narrow corridor down the middle. I don't even think they had a co-pilot – it was a pilot and a stewardess. So with the alcohol, things quite regularly got out of hand. Nine months I worked down there and there were two fights on the plane. Everybody from driller up was American, and everyone from derrickman down was Irish. One time this fight broke out between a little sub-sea engineer from Detroit and this huge Australian toolpusher. He was 6 ft 6 and this wee guy from Detroit, about 4 feet. They were scrapping in this little wee plane. The stewardess was crying. The pilot was trying to stop it. It was pandemonium. Another guy [once] set fire to his jacket. He had been smoking on the plane and he was drunk. Two or three times we were met in Aberdeen by the police.'

Sometimes the wayward oilmen had to pay for their unorthodox lifestyles. Ted Roberts, senior OIM on Forties, said one of the Americans was going out with a girl from Aberdeen. 'What we didn't know was that he was already married back in the States. One of our British lads found out and told this girl – she was a very nice girl – that this fellow was leading her on. I said to the American, "You haven't broken any rules as far as the company is concerned – but tell her." So he told her they were finished as he was already

THE OILMEN

married. Then he went back to work, but he had forgotten to get his key back from her. So she went in to his flat, dialled the New York time clock, and then left the phone off the hook. Then she dropped the key in the letterbox. When he returned ten days later he had one helluva phone bill.'

John Selbie was roustabouting for Shell and he was back loading containers on to a supply boat. 'I thought I would check everything was all right, so I opened the door and there were two mooring ropes – huge things, 4-inch diameter. I said to the roustie pusher, "This container is supposed to be empty but there are ropes in there." He said, "What are you doing opening that door? Shut it and just load them on. You never saw that." This was the kind of character you could be working with. That guy must have had a thing going with the guy on the beach. The ropes cost £400 pounds each, so he was getting £200 while his mate got the other £200.' John also saw a trailer being made in a welders' shop and then put on to a supply boat. 'Another night I went into the shop and there was this drill in the vice going flat out. It was plugged on to a speedo. The guy had taken the speedo off his motor and spun it right round the clock to get a low mileage on his car. That was going all night – in the welders' shop.'

Inside another workshop, the divers' workshop on the *Transworld 58* semi-submersible, Michel Euillet discovered a secret. 'I had been called on board to fix a television and I had to do some soldering. I unplugged something, which was marked 'do not unplug', to plug my soldering iron in. I followed the cable down and it led into some kind of pressure cooker. But I did my soldering and replugged it. It was actually a still for making alcohol and I had spoiled it. The alcohol was for the whole rig. I wasn't very popular.' Another time the French diver was on a Shell drilling rig, bossed by a Dutch toolpusher. 'We were looking at the BOP on television and we had been asked to run in reverse because they wanted to find a riser they had dropped. Well, we saw it and then when we panned down there were hundreds of beer cans on the seabed – Dutch beer. There wasn't supposed to be any alcohol on board. The toolpusher must have seen the pictures on his office monitor and immediately pulled the plug. As this guy was drinking he had been throwing the empties into the moon pool. Obviously the catering crew must have known for him to get his supplies of beer.'

The Neighbours

Oilmen, who spent their working lives on the steel islands offshore, were well aware of their neighbours in the form of the teeming marine life of the North

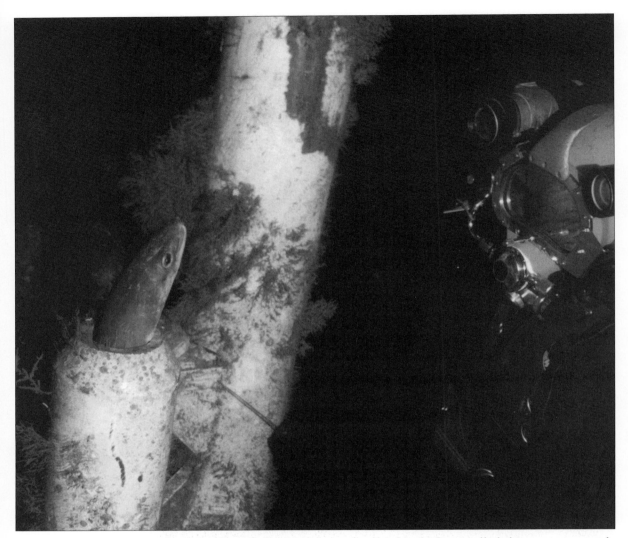

Plate 113.
Anyone for the conger? A giant eel pops out from his home in an oil platform pipe to greet a visitor. *(Kenny Thomson/ John Greensmyth, Technip Offshore)*

Sea as well as the huge variety of seabirds which patrolled the waters around them. The different species of fish were a constant diversion for the divers, while the amateur naturalists among the crews revelled in the wildlife – even forming a North Sea bird club which began on Forties and spread to other fields. On one occasion a cable was being laid along the seabed to Beryl Alpha. Roy Wilson of Mobil was in the diving boat. 'The sea was just like glass. It was a beautiful day, with little fish leaping around in the water. I was watching the divers working down below. Suddenly this whale appears and it came up right beside us – an amazing sight. It was fishing. You get a lot of fish around the platforms.' Another clear sunny day – this time in the Thistle field on the platform the oilmen call 'the Black Pig' – crewman Trevor Thorpe and a mate were standing on the main pipe deck, about 250 feet above sea level. 'My mate suddenly shouted, "What is that?" It was amazing.

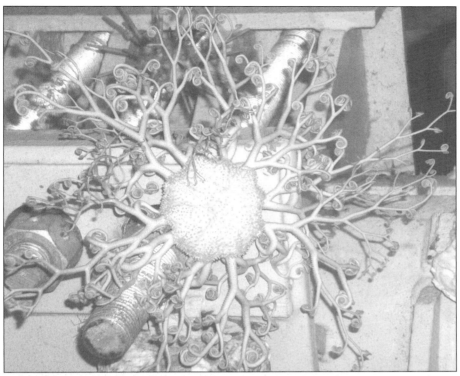

Plate 114 a and b.
Underwater neighbours for the oilmen – divers find a marine paradise around the artificial 'reefs' created by the oil and gas installations. *(Subsea 7)*

Plate 114 c and d.

THE OILMEN

The sea was almost white at the side of the rig. There must have been about a thousand pilot whales swimming past, heading north to their feeding grounds in the Arctic Ocean. Nobody had ever seen anything like it, except for an old guy who had worked on the trawlers. He told us it wasn't at all unusual.'

Trevor has a fund of nature tales. One concerned night shift on a very windy night. The Thistle had a permanent gas flare, which on a normal night pointed upwards, but then the wind was so strong it was almost horizontal. It was bright and the sky was clear. Trevor was working on the pipe deck when he saw large flocks of birds circling the rig. He couldn't tell what they were. He said the wind changed direction suddenly and the flare flipped back on itself. 'All we could hear were these soft thuds here and there. Looking around, we saw the deck was covered with little dark bodies. My mate looked at one and it was a starling, the next was a brambling, which is like a chaffinch. But the strangest of all the little bodies was that of a bat.' Trevor said he has often wondered what the bat was doing so far from land. 'The creatures had been caught out by the sudden shift in the wind, but they hadn't been burned: they had been struck down by the heat from the flare.'

A few years later Trevor was working on a semi-submersible, the *Ocean Liberator*, as a lead hand and assistant crane operator. The crew boss, Gary 'Bulldog' Jacobs, was teaching Trevor to lift containers from a supply boat. 'Suddenly he shouted. At the side of the supply boat a great whale had surfaced, taken a half roll and had a good look at the intruders. Strangely enough, the supply boat crew didn't see the whale and radioed to find out why we weren't lifting the next container.' Trevor added, 'I couldn't swear to it, but I got the feeling the whale was merely curious.' He experienced a number of wildlife visitors on that semi-sub, but the most amazing was when he and his mate, Ricky 'The Cat', were cleaning and checking the threads of some casing when they heard a cheeping noise. Sitting within inches of his nose was a goldcrest – the smallest British bird. 'I just couldn't believe it. I looked around me and realised they were all over the place, a huge flock using the rig as a resting place. I have never seen them on land, never mind at sea, but there were dozens of them. They really made my day and the rest of the two weeks just flew past.'

By far the strangest offshore intruder of all emerged out of a casing that had just been loaded on Brent Delta from a supply boat. Martin Reekie, who was on the crew, said the guy who opened it up got a real shock. 'Out staggered a hare. The casing had obviously been lying in a field somewhere and this creature managed to get into it. It was barely alive – it had been on the boat for three days. They tried to revive it but the thing died.'

THE OILMEN

The Crack

What most oilmen say they enjoyed most during their weeks offshore was the banter with their mates – the crack. 'It was brilliant,' said Jake Molloy. 'But then at that time there was no such thing as television. I never experienced anything like it again because it all went with the advent of television offshore. They all just go to their beds now to watch it. That was a good old time in the early days.' Mike Waller remembers an American from a casing cutting company coming down in the plane from Shetland. 'This diver was going on as they did. The American looked at him and said, "You always know a diver – he has a big watch and a small pecker."'

Peter Carson was an OIM on *Stadrill* and he had to oversea a paint crew one time. 'They were from Glasgow, the most wonderful people I ever worked with. It was my first real management of a crew. What humour! We had an old derrick which had rough spots on it and I was trying to get one guy to put some red lead primer around the monkey board. Well, the wind was blowing about 30 knots. He said, with suitable descriptive oaths, "Red lead primer? That's no' what you need. It's a squadron leader. I am no' goin' up there in that wind." We all laughed and I just walked away. But when I came back, he was getting his gear together and, of course, he was going up there. The crack was just so much more evident in those days.' One of Joe Cross's instructors was finishing a lecture and there was a little Glaswegian and a German sitting together. 'The German says to the Glaswegian, "And what do we do now?" The little guy comes back, "I don't know. Shall we invade Poland?"'

The Santa Fe manager from the shore, Bill Parry, an American in his sixties, went out regularly to West Sole A to see his drillers. BP engineer Jim Jenner said, 'When Bill came out, things jumped. He was doing his tour with one of the senior tool pushers and I was standing on the walkway on the pipe deck and Bill said hello. There were two roustabouts nearby, cleaning. Bill asked the toolpusher, "Are you short of any people on the floor?" A number of people had been quitting. He said, "Yes, I need another couple." Bill looked at the men cleaning, "Right, you two have been promoted, you are going to work on the floor." One of them said, "Oh, that's great." The other said, "I don't know. I am quite happy where I am." "OK," said Bill, "Go pack your bags. You are going ashore." As the guy walked away, he said, "Goddamn it. I can't stand people with no ambition."' That West Sole well was spudded-in by the American driller Gordon McCulloch. Jim Jenner recalled, 'We had landed the casing and we were down on the cellar deck, when a voice above said, "Heads!" You just automatically duck and hunch up, but the spinning chain – a length of about 15 feet, with 1-inch diameter links – had fallen through the rotary table. It hit Gordon across his hard hat and the end flipped and hit the point of his thumb – which he had lost on a previous occasion. He just stood there and looked at the end of his thumb. It hurt and he was shaking, but he just said, "Good job I had ma hard hat on."'

A delegation of Chinese dignitaries were visiting Ninian Central and the baker had been asked to make a cake with Chinese characters on it in their honour. When the cake was presented, everyone was really impressed, including the Chevron senior managers. But the leading Chinese laughed loudly, pointing at the cake. He asked who spoke Chinese as he was interested in the characters on the cake. What they actually said was 'as white as the driven snow'. The baker had to admit he had taken the Chinese words from a washing powder packet in the laundry.

The journalist who told the world about the first oil find, Ted Strachan, should have the last word in these North Sea tales. 'My wife was in Woolworths in Aberdeen one day and she heard two Aberdeen wifies speaking. One said her man had gone offshore. "He's awa' in Las Palmas." This was in the depths of winter. Then she added, "It's aye a job, intit?"'

Epilogue

Old Age or a New Age?

'Please yersel' laddie. You ging affshore, bit the ile will seen ging awa'.
An' you'll be needin' yer job back.'
Advice from an Aberdeen employer to an aspiring oilman in 1975

The sale of BP's talismanic oilfield, Forties, to a new entrant company in the North Sea appeared at first to be symbolic of the decline of the industry that has flourished on the UK Continental Shelf. Just as the discovery signalled the beginning of the country's most important large-scale enterprise of the twentieth century, the £401 million transfer of ownership was suspiciously like a portent of the beginning of the end, with the world's second-biggest energy corporation deserting the arena that had made them great.

That pessimistic interpretation has been further reinforced by the wholesale divestment of a bundle of assets, interests and some exploration acreage by the other giant, Shell. In other areas, ominous numbers flag up a slow but steady demise. The output of oil and gas is falling perceptibly and is unlikely to increase to the levels of the peak year of 1999. There has also been a lengthy hiatus in exploration and drilling – admittedly a global problem – resulting in queues of idle rigs, twenty-four at one stage in the Cromarty Firth, and a shrinking work force among contractors and service companies. While some drilling companies are optimistically forecasting an upturn, a Trade and Industry survey has shown that general expectations are low. Offshore, a number of fields have either dried up or become uneconomic, while twenty-five of the 250 offshore fixed and mobile installations have been uprooted since 1988 and, although some fields will be reworked, the majority of the reservoirs have been abandoned. All the big oil traps are thought now to have been identified. The surge of innovative technology that surmounted the unique problems of the province seems to

Figure 6.
The empty wastes of the North Sea, now crowded with all the trappings of the multi-billion pound UK oil and gas industry. *(Greybardesign)*

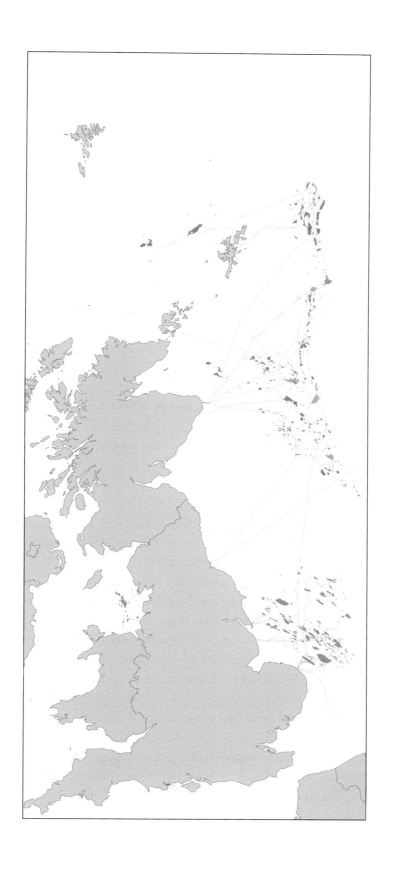

THE OILMEN

have faltered and is awaiting investment. West of Shetland on the North Atlantic margins, further expansion beyond Foinaven, Schiehallion and Clair appears to have stalled. The UK oil and gas industry, it is said, has achieved maturity and the great beasts of the global petroleum and gas markets are slouching off and are now fully engrossed in redirecting their money and talents towards cheaper developing countries, while plundering the indigent treasure trove of oilfield skills nurtured over the years to exploit these burgeoning provinces in Africa, Asia and South America.

Nevertheless, against this apparent gathering of black clouds, the financial and statistical report card shows that the North Sea remains a prized national asset of strategic value to the country. Since the start of development in the southern gas fields in 1965, the industry has generated £306 billion and reinvested some £210 billion. The contribution to industrial investment in the UK has been around 17 per cent, more than 2 per cent of gross value added. The direct impact of production has truly propped up the balance of payments – with a contribution soaring to £8 billion in 1985, although it has slipped in 2004 to nearer £5.5 billion. There has also been a steady flow of revenue in taxes and royalties, again peaking in 1985 at £11 billion and with an overall total currently standing at more than £100 billion. Another major factor has been in the national labour market. The present UK total on and offshore of oil-related and oil-dependent employment is estimated at 265,000; of these 31 per cent are in Scotland, representing 6 per cent of the Scottish workforce. By any industrial or economic standards these figures, like the industry itself, are epic (see appendices).

Over the past four decades the empty seas off the Scottish coast have undergone an unimaginable transformation: from the lonely, tentative exploration by the semi-submersible sisters *Sea Quest* and *Staflo* across the central and northern waters, and, further east, by the *Ocean Viking*, to around 150 fields with their colonies of platforms and drilling rigs, thirled to thousands of miles of pipeline systems leading to four landfalls. Onshore, too, whole oil communities and service centres and bases have prospered. None of the bold pioneers who described how they expanded the frontiers of technology in a number of marine oil disciplines could have foreseen the growth of such a bustling, tireless industry or that it would be the purveyor of such great wealth both nationally and locally, and provide so much employment. Very few of them would have believed that the juggernaut they had set in motion would outrun their working lives and beyond. And still they can't.

But there are those inside and outside the oil and gas business who do believe it. They are firmly convinced the industry will continue to run and that it is at present basking in healthy middle age and nowhere near industrial senility. They argue that it is, in fact, poised on the brink of something new – a new life, a new age, or era, or more modestly, a new phase in the continuing story of the North Sea.

What seems certain, however, is that the UKCS has reached a crossroads and the companies, new and old, now find themselves like the original small firms, bold explorers contemplating the unknown. Just as the industry was forced after its worst ever disaster to confront the fatal flaws in what was considered to be a safe working environment, oilmen must look back into their own history and retrieve the imperatives which inspired those who first went north across the 56th parallel. There really are 'lessons to be learned'. There is another issue, however. With a few notable exceptions, sceptical British industry and British entrepreneurs largely missed out on the opportunities fortuitously presented to them – certainly in manufacturing and most definitely in operatorships. Now the knowledge and experience unavailable thirty years ago is stored in the nation's memory banks. Whatever the next advance, it represents a second chance for UK industry, for Scottish and British investors and entrepreuners, and for the politicians.

The bold road forward is undoubtedly littered with many unresolved issues; there is the undeveloped acreage held by the major companies – estimated at some 50 per cent of reserves. The unions – and the new companies – argue that if BP and Shell intend to concentrate on the new provinces overseas, they should release the unexplored blocks on the UKCS. One of the new men, Paul Blakeley of the biggest independent, Talisman, claims that as much as 10 billion barrels of oil could remain untapped because the big companies are no longer interested in extracting the hydrocarbons. Both BP and Shell maintain, however, that they are continuing to invest hugely in developing their remaining assets. BP, for example, have now committed to new, larger headquarters in Aberdeen. Another that is that these majors still retain control of the pipeline systems transporting the oil and gas. Access is therefore available only by rental. BP again have acted by reducing the cost of hiring the use of the Forties system by a third.

Then there is the political involvement. Despite the participation of the industry in Government-inspired initiatives such as PILOT, the Chancellor of the Exchequer's unexpected 10 per cent tax on profits in the 2002 Budget damaged the relationship the shocked industry believed had been established with Whitehall. The Government has since attempted to make amends with measures including the abolition of royalty and petroleum revenue tax on new tarriff busines and the introduction of a supplement on exploration expenditure. Another attempted filip to encourage development boost came

THE OILMEN

in the twenty-first block allocation with the new Promote licence, which eases the way in for smaller newcomers. But the industry – particularly the drilling sector – wants further tax incentives to enable them to pursue new prospects.

A further barrier to the long-term future is the slower than expected introduction of the enhanced technology needed for progress. Innovators complain bitterly of the reluctance of the industry to invest time and money on new ideas. They point to the excessively long lead times before their new concepts are accepted and developed commercially. Charlie Anderton, whose company specialises in new downhole technology, explained: 'First of all the oil companies don't want the hassle of making products – they just want to rent them, use them and give them back. Otherwise they have to carry a huge infrastructure. Research and development is expensive. The cheapest way is to use small companies like mine. My first year I sold directly to the oil companies. Then they brought out the "win-win" scheme using sole contractors, and all the other companies had to go through the one-service supplier who takes his commission off them. If your technology competes with theirs, then they won't let you in. So now there are fewer innovative firms and the oil companies don't seem to realise they are strangling innovation.' Oil industry commentator Jeremy Cresswell has argued that new entrant minnows with less money to spend may be less inclined to indulge in risk-technology which would threaten the advance of oilfield knowledge and techniques in the longer term.

Other problems concern the severe skill shortage which exists at a time when the current offshore workforce is ageing. As a number of the women interviewed indicated, recruitment of a steady stream of young people is proving difficult in an industry with an average age of forty-five to fifty in what was a young man's game. Without a supply of willing new pupils, the skills, hard earned by UK oil workers nearing retirement, will vanish.

The industry, competing aggressively for talent with financial services and information technology have launched a number of skills-learning initiatives, including one for subsea technology, involving universities, contractors and unions, while the Government are running a training network called Cogent. Alison Golligher of Schlumberger said that one of the difficulties was possibly the 'bad' image the oil and gas industry has. 'I think it is a wee bit of a shame, because this industry has done a lot for the UK, putting a lot of money in the coffers. I think it provides a great opportunity for youngsters in an area with a huge amount of specialisation and where they can see the world. It is a hugely adventurous and challenging business. Sometimes I think we should stop talking about how difficult it is to attract people in and start talking about how easy it is.'

What adds credence to the argument about the prospect of new life for an old province is the fact that nearly half as much again oil and gas has still to be recovered. The amounts are the subject of debate – 20 billion to 30 billion barrels are said to be waiting in the reservoirs. The OCA's Bill Murray claimed, 'To some extent what has been pulled out is just the low-hanging fruit. The easy stuff to bring up. We haven't really started to explore the technologies that will bring forward the more difficult prizes. There is probably as much heavy crude out there – I have heard mention of 14 or 15 billion barrels. It is a costly challenge, admittedly, but there are far better opportunities in the North Sea, particularly for the smaller operators. You have a stable economic environment, a stable political climate and a workforce, well-educated, highly skilled, able to adapt.'

An illustration of companies choosing the easy option in the rush to extract oil, comes from Alan Higgins, former North Sea General Manager at Chevron. 'They did take the easy way, and as a result they didn't actually manage the fields terribly well. A new company can come in now and recover more. When we started, at best, we got back about 30 to 35 per cent of the oil. Now some fields are up to 50 and 60 per cent with improved technology, better management and better placing of the wells to recover the maximum amount of oil. So there is all that oil left in the existing fields if you can but get the technology.'

For many, the first flickers of a revival in the North Sea have already appeared with the arrival of the small and medium-sized companies known as 'new entrants'. One company's divestment is another's investment and for them, the offloading of assets by big players such as Shell and BP opens the door to great new opportunities. These incomers are either buying ageing fields, such as Apache with Forties, or investing in co-ownerships, backed by venture capital. In the last licensing round, twenty-seven new companies were attracted through the new Promote licence. Government agencies are also making a big push to woo independent US and Canadian exploration and production companies. The new men are exploring new possibilities in energising the industry. One imaginative financial intiative is being pioneered by Energy Development Partners (EDP), who have have set up a $100-million fund available to global investors. Their first move was to broker a deal with BP to work two of the company's southern oilfields, Wareham and Kimmeridge, earning a share of production from the giant.

A new entrant who can visualise the opportunities in Bill Murray's unpicked 'fruit' is John Crum, North Sea executive vice-president of Apache. His company are engaged in a £43-million investment programme on Forties

THE OILMEN

and plan to push production back up to 70 per cent. But the real shock new entrant is the Canadian company EnCana, who discovered the largest new field in the North Sea in the past ten years. The find, Buzzard, lies sixty miles north-east of Aberdeen and is thought to contain 1.2 billion barrels of oil and total recoverable reserves of more than 400 million. And this at a time when the industry had believed there were no more 'elephants' lurking in the UKCS. EnCana, who have a 43 per cent interest, are in the throes of a £1.35-billion development programme and the field is expected to start producing in 2005. Energy Minister Steven Timms, who has been busy talking up the North Sea as a viable prime market for overseas companies, claims Buzzard is proof that the province is still vibrant and rewarding. 'I hope it will act as a beacon to attract new entrants and renewed commitment from existing players.' But realistically the input by the new entrants will not be enough – any significant North Sea revival still depends on the massive financial clout of the majors.

So the stage is set, and in the nature of the industry there has been a mesmeric regrouping of older cast members as the new players assemble. 'I think we will be in the situation the industry was in at the start,' said Bill Murray. 'The will is there. There is a huge global prize for companies to do this. We can also offer expertise as we are now doing with the Norwegians, which is partly the prize for them selling us their gas.' This is the agreement signed with the UK's gas-rich neighbours to safeguard declining reserves of the energy form.

There remain other issues to settle. It is extraordinary that after nearly forty years the industry has still not come to terms with the merits of modern trade unions. If companies honestly wish to comunicate with their workforce and estabish the understanding needed to build a safe environment offshore, the unions are the obvious route. Similarly, the official trade union movement has to bury its historic differences with the OILC and capitalise on the strengths of a genuine grass-roots movement largely trusted by the offshore workforce.

At the heart of it all is the fundamental question: do oilmen of the twenty-first century possess the spirit, the innovative talent, the nerve to gamble huge sums and the drive to rejuvenate the North Sea? Just as in the 1970s, when oil and gas were described as 'God's last chance for the British', it will take the efforts of everyone with a stake in the industry – operators, contractors, workforce and Government. Maybe this is their second chance. For history's greatest 'can-do, problem-solving, need-it-yesterday' breed the challenge is there. The early American oil boss, Jack Marshall lived in a time when they were given 'the centre cut of the water melon'. For oilmen who have the will, that still leaves most of the fruit to savour.

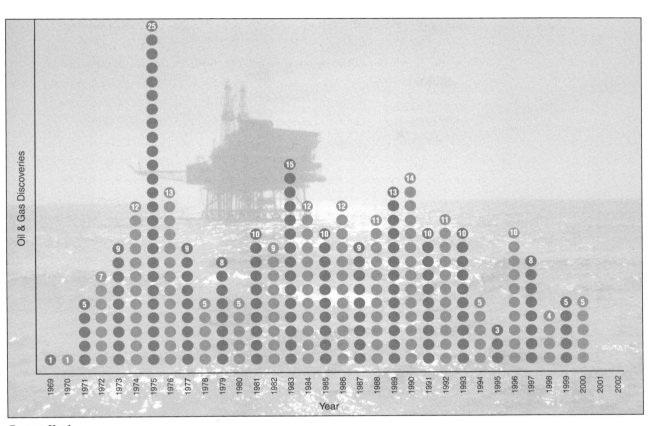

Appendix 1.
Thirty years of discoveries in the North Sea oil and gas industry. *(Greybardesign)*

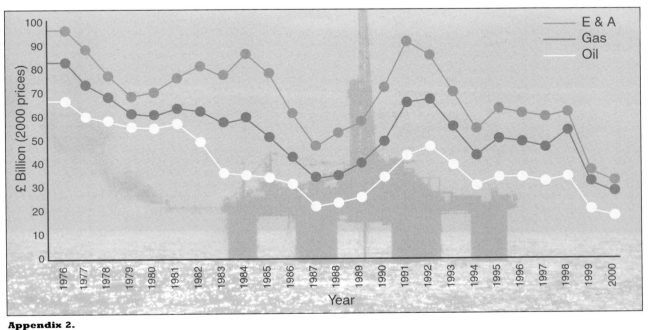

Appendix 2.
The rate of expenditure. *(Greybardesign)*

[259]

Appendix 3.

The rise and fall in employment. *(Greybardesign)*

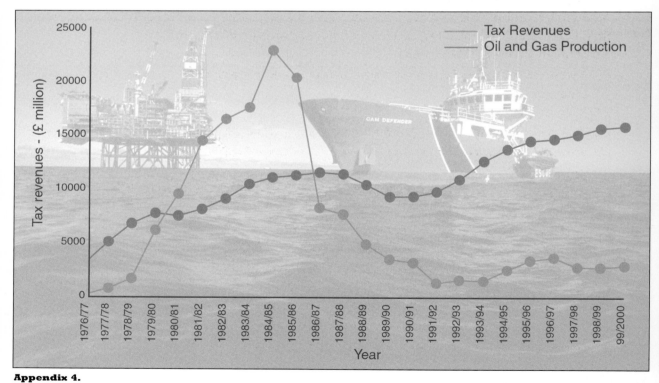

Appendix 4.

Revenues from the oil and gas taxation and annual production. *(Greybardesign)*

Glossary

Abandonment Converting a well to be left indefinitely.
Amicus Main North Sea oil union formed by merger of the engineering union, the AEU and MSF, the scientific, technical and management union.
Annulus Space between two pipes or between a pipe and the well bore.
Atlantic Margin Largely unexplored acreage in the Atlantic to the West of Shetland, believed to be a fertile source of hydrocarbons.
BALPA British Airline Pilots' Association.
Barrel The unit of oil and gas volume measurement – 7.3 barrels equal 1 ton; 6.29 equal 1 cubic metre.
Blow out Uncontrolled flow of oil, gas or fluids from a well.
BNOC The British National Oil Corporation, the state-owned oil company set up in 1976 and abolished in 1985.
BOP Blow out preventer – equipment installed at the wellhead to control pressure.
BPD The number of barrels per day.
Casing Steel lining in the bore to control pressures and support the sides.
Christmas tree A collection of pipes, gauges and valves controlling oil and gas flow.
CNS Central North Sea area of the UKCS. The other sectors are the northern North Sea, the Moray Firth and the southern North Sea.
Completion Finishing a well ready to produce oil and gas.
Condensate Liquid mixture of lighter and higher hydrocarbons.
CRINE Cost Reduction in a New Era – oil industry and government economic initiative, replaced by *PILOT*.
Cuttings Fragments of rock brought to the surface by the drilling mud.
Decommisioning Dismantling and removing an offshore installation.
Derrick Steel pyramid mounted over the bore hole for suspending and rotating drill pipes.
Drill string Connected drill pipes or joints.

Downstream Generally the refining and marketing sectors.

Directional drilling Deviation of the drill from the vertical to explore the extent of a field.

Drilling mud Mixture with additives used to lubricate drilling and counteract natural pressures. Also provides analysis of the state of a well.

ELSBM exposed location buoy.

Flaring and flare stack Burning of gases for commercial or technical reasons through a stack.

Geophysics The science of the interaction between the physical features of the Earth and the forces that produced them – including seismology and magnetism.

Hydrocarbons Coal, crude oil and natural gas, composed of carbon and hydrogen

Jacket Supporting structure for an installation

Kick Entry of higher pressure fluids into the wellbore with the potential to cause a blow out.

Kelly Square or hexagonal steel tube fixed to the drill rotary table.

Licensing round The periodic awards of licenses to the oil companies and interested parties for blocks of North Sea acreage. The exploitation of the blocks are subject to time restrictions.

LNG Liquefied natural gas.

LPG Liquefied petroleum gas.

Marine riser Pipe connecting an installation to a wellhead or pipeline.

Module Self-contained structure located on installations with a variety of production and operational uses.

Moonpool Drill hole in a rig or semi-submersible.

NUS National union of seamen.

OCA Offshore contractors' association – a trade organisation with official industrial relations and wage bargaining powers.

OILC Offshore industry liasion committee. The unofficial organisation founded by the Bears, the construction workers, after the *Piper Alpha* disaster. Now an official union.

OIM Offshore installations manager – in charge of platform operations.

PILOT A cross-industry and Government task force.

PRT Petroleum Revenue Tax, introduced in 1975, and applies to UK oil production and profits.

Rathole (mousehole) Opening on the drill floor for the kelly.

Roughneck Drilling crew member.

Roustabout Drill floor staff who handle loading and general operational tasks.

SBM Single point buoy mooring for loading.

Seismic survey Measurement of the effects of subsea seismic waves created by the firing of shots either by explosives or gas guns. Current technology includes 3 and 4 dimensional seismology which is computer generated.

Shale shaker Used to sieve out drill cuttings from the mud fluids.

Shutdown Maintenance and platform service break (usually during the summer months).

Spar A manoeuverable loading vessel.

Spider deck The platform occupied by the derrickman.

SPM Single point mooring

Stack A steel structure on an installation.

Step Change Oil industry/Government/union iniative to introduce a safety regime.

Stinger A curved gantry down which pipe is fed, on a pipe-laying barge.

Subsea wellhead Seabed well production installation on the seabed and remotely controlled.

TLP Tension leg platform – floating offshore structure held in place by cable anchors.

Toolpusher Second in command on a drillling rig under the superintendent

TGWU Transport and General Workers' Union.

UCM Underwater central manifold – seabed well production installation that doesn't need a platform.

UKCS United Kingdom Continental Shelf – the extent of the UK's territorial licensing areas in the North Sea which came into force under the Continental Shelf Act in 1964.

UKOA The United Kingdom Offshore Operators' Association – the oil companies' lobby group.

Wildcat well Well drilled in virgin area.